Data Analysis
for the
Life Sciences
with R

Data Analysis for the Life Sciences with R

Rafael A. Irizarry

Michael I. Love

CRC Press
Taylor & Francis Group
Boca Raton London New York

CRC Press is an imprint of the
Taylor & Francis Group, an **informa** business

A CHAPMAN & HALL BOOK

CRC Press
Taylor & Francis Group
6000 Broken Sound Parkway NW, Suite 300
Boca Raton, FL 33487-2742

© 2017 by Taylor & Francis Group, LLC
CRC Press is an imprint of Taylor & Francis Group, an Informa business

No claim to original U.S. Government works

Printed on acid-free paper
Version Date: 20160614

International Standard Book Number-13: 978-1-4987-7567-0 (Hardback)

Visit the Taylor & Francis Web site at
http://www.taylorandfrancis.com

and the CRC Press Web site at
http://www.crcpress.com

Printed in Canada

Contents

List of Figures

Acknowledgments

The authors would like to thank Alex Nones for proofreading the manuscript during its various stages. Also, thanks to Karl Broman for contributing the "Plots to Avoid" section and to Stephanie Hicks for designing some of the exercises. Finally, thanks to John Kimmel and three anonymous referees for excellent feedback and constructive criticism of the book.

This book was conceived during the teaching of several HarvardX courses, coordinated by Heather Sternshein. We are also grateful to our TAs, Idan Ginsburg and Stephanie Chan, and all the students whose questions and comments helped us improve the book. The courses were partially funded by NIH grant R25GM114818. We are very grateful to the National Institute of Health for its support.

A special thanks goes to all those that edited the book via GitHub pull requests: vjcitn, yeredh, ste-fan, molx, kern3020, josemrecio, hcorrada, neerajt, massie, jmgore75, molecules, lzamparo, eronisko, obicke, knbknb, and devrajoh.

Cover image credit: this photograph is La Mina Falls, El Yunque National Forest, Puerto Rico, taken by Ron Kroetz https://www.flickr.com/photos/ronkroetz/14779273923 Attribution-NoDerivs 2.0 Generic (CC BY-ND 2.0)

Introduction

The unprecedented advance in digital technology during the second half of the 20th century has produced a measurement revolution that is transforming science. In the life sciences, data analysis is now part of practically every research project. Genomics, in particular, is being driven by new measurement technologies that permit us to observe certain molecular entities for the first time. These observations are leading to discoveries analogous to identifying microorganisms and other breakthroughs permitted by the invention of the microscope. Choice examples of these technologies are microarrays and next generation sequencing.

Scientific fields that have traditionally relied upon simple data analysis techniques have been turned on their heads by these technologies. In the past, for example, researchers would measure the transcription levels of a single gene of interest. Today, it is possible to measure all 20,000+ human genes at once. Advances such as these have brought about a shift from hypothesis to discovery-driven research. However, interpreting information extracted from these massive and complex datasets requires sophisticated statistical skills as one can easily be fooled by patterns arising by chance. This has greatly elevated the importance of statistics and data analysis in the life sciences.

Who Will Find This Book Useful?

This book was written with the many life science researchers who are becoming data analysts due to the increased reliance on data described above. If you are performing your own analysis you have probably computed p-values, applied Bonferroni corrections, performed principal component analysis, made a heatmap, or used one or more of the techniques listed in the next section. If you don't quite understand what these techniques are actually doing or if you are not sure if you are using them appropriately, this book is for you.

Although the content of the book is mostly focused on advanced statistical concepts we start by covering the basics to make sure all readers have a strong grounding on the fundamental statistical concepts required for all data analysis. I find that many introductory statistics courses are taught in a way that makes it hard to relate the concepts to data analysis. Our approach ensures that you learn the connection between practice and theory. For this reason, the first two chapters, Inference and Exploratory Data Analysis, are appropriate for an introductory undergraduate statistics or data science course. After these two chapters the level of statistical sophistication ramps up relatively fast.

Although the typical reader of this book will have a masters or PhD, we try to keep the mathematical content at undergraduate introductory level. You do not need calculus to use this book. However, we do introduce and use linear algebra which is considered more advanced than calculus. By explaining linear algebra in context of data analysis we believe you will be able to learn the basics without knowing calculus. The harder part may be getting used to the symbols and notation. More on this below.

What Does This Book Cover?

This book will cover several of the statistical concepts and data analytic skills needed to succeed in data-driven life science research. We go from relatively basic concepts related to computing p-values to advanced topics related to analyzing high-throughput data.

We start with one of the most important topics in statistics and in the life sciences: statistical inference. Inference is the use of probability to learn population characteristics from data. A typical example is deciphering if two groups (for example, cases versus controls) are different on average. Specific topics covered include the t-test, confidence intervals, association tests, Monte Carlo methods, permutation tests and statistical power. We make use of approximations made possible by mathematical theory, such as the Central Limit Theorem, as well as techniques made possible by modern computing. We will learn how to compute p-values and confidence intervals and implement basic data analyses. Throughout the book we will describe visualization techniques in the statistical computer language R that are useful for exploring new datasets. For example, we will use these to learn when to apply robust statistical techniques.

We will then move on to an introduction to linear models and matrix algebra. We will explain why it is beneficial to use linear models to analyze differences across groups, and why matrices are useful to represent and implement linear models. We continue with a review of matrix algebra, including matrix notation and how to multiply matrices (both on paper and in R). We will then apply what we covered on matrix algebra to linear models. We will learn how to fit linear models in R, how to test the significance of differences, and how the standard errors for differences are estimated. Furthermore, we will review some practical issues with fitting linear models, including collinearity and confounding. Finally, we will learn how to fit complex models, including interaction terms, how to contrast multiple terms in R, and the powerful technique which the functions in R actually use to stably fit linear models: the QR decomposition.

In the third part of the book we cover topics related to high-dimensional data. Specifically, we describe multiple testing, error rate controlling procedures, exploratory data analysis for high-throughput data, p-value corrections and the false discovery rate. From here we move on to covering statistical modeling. In particular, we will discuss parametric distributions, including binomial and gamma distributions. Next, we will cover maximum likelihood estimation. Finally, we will discuss hierarchical models and empirical Bayes techniques and how they are applied in genomics.

We then cover the concepts of distance and dimension reduction. We will introduce the mathematical definition of distance and use this to motivate the singular value decomposition (SVD) for dimension reduction and multi-dimensional scaling. Once we learn this, we will be ready to cover hierarchical and k-means clustering. We will follow this with a basic introduction to machine learning.

We end by learning about batch effects and how component and factor analysis are used to deal with this challenge. In particular, we will examine confounding, show examples of batch effects, make the connection to factor analysis, and describe surrogate variable analysis.

How Is This Book Different?

While statistics textbooks focus on mathematics, this book focuses on using a computer to perform data analysis. This book follows the approach of Stat Labs[1], by Deborah Nolan and Terry Speed. Instead of explaining the mathematics and theory, and then showing examples,

---

[1]https://www.stat.berkeley.edu/~statlabs/

we start by stating a practical data-related challenge. This book also includes the computer code that provides a solution to the problem and helps illustrate the concepts behind the solution. By running the code yourself, and seeing data generation and analysis happen live, you will get a better intuition for the concepts, the mathematics, and the theory.

We focus on the practical challenges faced by data analysts in the life sciences and introduce mathematics as a tool that can help us achieve scientific goals. Furthermore, throughout the book we show the R code that performs this analysis and connect the lines of code to the statistical and mathematical concepts we explain. All sections of this book are reproducible as they were made using *R markdown* documents that include R code used to produce the figures, tables and results shown in the book. In order to distinguish it, the code is shown in the following font:

```
x <- 2
y <- 3
print(x+y)
```

and the results in different colors, preceded by two hash characters (*##*):

```
## [1] 5
```

We will provide links that will give you access to the raw R markdown code so you can easily follow along with the book by programming in R.

At the beginning of each chapter you will see the sentence:

The R markdown document for this section is available here.

The word "here" will be a hyperlink to the R markdown file. The best way to read this book is with a computer in front of you, scrolling through that file, and running the R code that produces the results included in the book section you are reading.

1

Getting Started

In this book we will be using the R programming language[1] for all our analysis. You will learn R and statistics simultaneously. However, we assume you have some basic programming skills and knowledge of R syntax. If you don't, your first homework, listed below, is to complete a tutorial. Here we give step-by-step instructions on how to get set up to follow along.

1.1 Installing R

The first step is to install R. You can download and install R from the Comprehensive R Archive Network[2] (CRAN). It is relatively straightforward, but if you need further help you can try the following resources:

- Installing R on Windows[3]
- Installing R on Mac[4]
- Installing R on Ubuntu[5]

1.2 Installing RStudio

The next step is to install RStudio, a program for viewing and running R scripts. Technically you can run all the code shown here without installing RStudio, but we highly recommend this integrated development environment (IDE). Instructions are here[6] and for Windows we have special instructions[7].

1.3 Learn R Basics

The first homework assignment is to complete an R tutorial to familiarize yourself with the basics of programming and R syntax. To follow this book you should be familiar with

---

[1]https://cran.r-project.org/
[2]https://cran.r-project.org/
[3]https://github.com/genomicsclass/windows#installing-r
[4]http://youtu.be/Icawuhf0Yqo
[5]http://cran.r-project.org/bin/linux/ubuntu/README
[6]http://www.rstudio.com/products/rstudio/download/
[7]https://github.com/genomicsclass/windows

the difference between lists (including data frames) and numeric vectors, for-loops, how to create functions, and how to use the `sapply` and `replicate` functions.

If you are already familiar with R, you can skip to the next section. Otherwise, you should go through the swirl[8] tutorial, which teaches you R programming and data science interactively, at your own pace and in the R console. Once you have R installed, you can install `swirl` and run it the following way:

```
install.packages("swirl")
library(swirl)
swirl()
```

Alternatively you can take the try R[9] interactive class from Code School.

There are also many open and free resources and reference guides for R. Two examples are:

- Quick-R[10]: a quick online reference for data input, basic statistics and plots
- R reference card (PDF)[https://cran.r-project.org/doc/contrib/Short-refcard.pdf] by Tom Short

Two key things you need to know about R is that you can get help for a function using `help` or ?, like this:

```
?install.packages
help("install.packages")
```

and the hash character represents comments, so text following these characters is not interpreted:

```
##This is just a comment
```

1.4   Installing Packages

The first R command we will run is `install.packages`. If you took the `swirl` tutorial you should have already done this. R only includes a basic set of functions. It can do much more than this, but not everybody needs everything so we instead make some functions available via packages. Many of these functions are stored in CRAN. Note that these packages are vetted: they are checked for common errors and they must have a dedicated maintainer. You can easily install packages from within R if you know the name of the packages. As an example, we are going to install the package `rafalib` which we use in our first data analysis examples:

```
install.packages("rafalib")
```

We can then load the package into our R sessions using the `library` function:

```
library(rafalib)
```

From now on you will see that we sometimes load packages without installing them. This is because once you install the package, it remains in place and only needs to be loaded with `library`. If you try to load a package and get an error, it probably means you need to install it first.

---

[8]http://swirlstats.com/
[9]http://tryr.codeschool.com/
[10]http://www.statmethods.net/

1.5   Importing Data into R

The first step when preparing to analyze data is to read in the data into R. There are several ways to do this and we will discuss three of them. But you only need to learn one to follow along.

In the life sciences, small datasets such as the one used as an example in the next sections are typically stored as Excel files. Although there are R packages designed to read Excel (xls) format, you generally want to avoid this and save files as comma delimited (Comma-Separated Value/CSV) or tab delimited (Tab-Separated Value/TSV/TXT) files. These plain-text formats are often easier for sharing data with collaborators, as commercial software is not required for viewing or working with the data. We will start with a simple example dataset containing female mouse weights[11].

The first step is to find the file containing your data and know its *path*.

Paths and the Working Directory   When you are working in R it is useful to know your *working directory*. This is the directory or folder in which R will save or look for files by default. You can see your working directory by typing:

```
getwd()
```

You can also change your working directory using the function `setwd`. Or you can change it through RStudio by clicking on "Session".

The functions that read and write files (there are several in R) assume you mean to look for files or write files in the working directory. Our recommended approach for beginners will have you reading and writing to the working directory. However, you can also type the full path[12], which will work independently of the working directory.

Projects in RStudio   We find that the simplest way to organize yourself is to start a Project in RStudio (Click on "File" and "New Project"). When creating the project, you will select a folder to be associated with it. You can then download all your data into this folder. Your working directory will be this folder.

Option 1: Read file over the Internet   You can navigate to the `femaleMiceWeights.csv` file by visiting the data directory of dagdata on GitHub[13]. If you navigate to the file, you need to click on *Raw* on the upper right hand corner of the page.

Now you can copy and paste the URL and use this as the argument to `read.csv`. Here we break the URL into a base directory and a filename and then combine with `paste0` because the URL would otherwise be too long for the page. We use `paste0` because we want to put the strings together as is, if you were specifying a file on your machine you should use the smarter function, `file.path`, which knows the difference between Windows and Mac file path connectors. You can specify the URL using a single string to avoid this extra step.

```
dir <- "https://raw.githubusercontent.com/genomicsclass/dagdata/master/inst/extdata/"
url <- paste0(dir, "femaleMiceWeights.csv")
dat <- read.csv(url)
```

---

[11]https://raw.githubusercontent.com/genomicsclass/dagdata/master/inst/extdata/femaleMiceWeights.csv

[12]http://www.computerhope.com/jargon/a/absopath.htm

[13]https://github.com/genomicsclass/dagdata/tree/master/inst/extdata

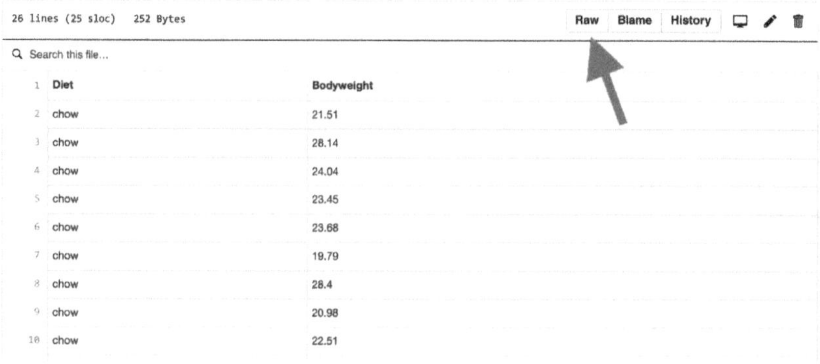

26 lines (25 sloc) 252 Bytes		Raw Blame History ⬚ ✎ 🗑
🔍 Search this file...		
1 **Diet**	**Bodyweight**	
2 chow	21.51	
3 chow	28.14	
4 chow	24.04	
5 chow	23.45	
6 chow	23.68	
7 chow	19.79	
8 chow	28.4	
9 chow	20.98	
10 chow	22.51	

FIGURE 1.1
GitHub page screenshot.

Option 2: Download file with your browser to your working directory There are reasons for wanting to keep a local copy of the file. For example, you may want to run the analysis while not connected to the Internet or you may want to ensure reproducibility regardless of the file being available on the original site. To download the file, as in option 1, you can navigate to the `femaleMiceWeights.csv`. In this option we use your browser's "Save As" function to ensure that the downloaded file is in a CSV format. Some browsers add an extra suffix to your filename by default. You do not want this. You want your file to be named `femaleMiceWeights.csv`. Once you have this file in your working directory, then you can simply read it in like this:

```
dat <- read.csv("femaleMiceWeights.csv")
```

If you did not receive any message, then you probably read in the file successfully.

Option 3: Download the file from within R We store many of the datasets used here on GitHub[14]. You can save these files directly from the Internet to your computer using R. In this example, we are using the `download.file` function in the `downloader` package to download the file to a specific location and then read it in. We can assign it a random name and a random directory using the function `tempfile`, but you can also save it in directory with the name of your choosing.

```
library(downloader) ##use install.packages to install
dir <- "https://raw.githubusercontent.com/genomicsclass/dagdata/master/inst/extdata/"
filename <- "femaleMiceWeights.csv"
url <- paste0(dir, filename)
if (!file.exists(filename)) download(url, destfile=filename)
```

We can then proceed as in option 2:

```
dat <- read.csv(filename)
```

Option 4: Download the data package (Advanced) Many of the datasets we include in this book are available in custom-built packages from GitHub. The reason we use GitHub, rather than CRAN, is that on GitHub we do not have to vet packages, which gives us much more flexibility.

---

[14]https://github.com/genomicsclass/

To install packages from GitHub you will need to install the `devtools` package:

```
install.packages("devtools")
```

Note to Windows users: to use devtools you will have to also install `Rtools`. In general you will need to install packages as administrator. One way to do this is to start R as administrator. If you do not have permission to do this, then it is a bit more complicated[15].

Now you are ready to install a package from GitHub. For this we use a different function:

```
library(devtools)
install_github("genomicsclass/dagdata")
```

The file we are working with is actually included in this package. Once you install the package, the file is on your computer. However, finding it requires advanced knowledge. Here are the lines of code:

```
dir <- system.file(package="dagdata") #extracts the location of package
list.files(dir)
```

```
## [1] "data"        "DESCRIPTION" "extdata"     "help"        "html"
## [6] "Meta"        "NAMESPACE"   "script"
```

```
list.files(file.path(dir,"extdata")) #external data is in this directory
```

```
## [1] "admissions.csv"              "babies.txt"
## [3] "femaleControlsPopulation.csv" "femaleMiceWeights.csv"
## [5] "mice_pheno.csv"              "msleep_ggplot2.csv"
## [7] "README"                      "spider_wolff_gorb_2013.csv"
```

And now we are ready to read in the file:

```
filename <- file.path(dir,"extdata/femaleMiceWeights.csv")
dat <- read.csv(filename)
```

1.6   Exercises

Here we will test some of the basics of R data manipulation which you should know or should have learned by following the tutorials above. You will need to have the file `femaleMiceWeights.csv` in your working directory. As we showed above, one way to do this is by using the `downloader` package:

```
library(downloader)
dir <- "https://raw.githubusercontent.com/genomicsclass/dagdata/master/inst/extdata/"
filename <- "femaleMiceWeights.csv"
url <- paste0(dir, filename)
if (!file.exists(filename)) download(url, destfile=filename)
```

1. Read in the file `femaleMiceWeights.csv` and report the body weight of the mouse in the exact name of the column containing the weights.

---

[15]http://www.magesblog.com/2012/04/installing-r-packages-without-admin.html

2. The [and] symbols can be used to extract specific rows and specific columns of the table. What is the entry in the 12th row and second column?

3. You should have learned how to use the $ character to extract a column from a table and return it as a vector. Use $ to extract the weight column and report the weight of the mouse in the 11th row.

4. The `length` function returns the number of elements in a vector. How many mice are included in our dataset?

5. To create a vector with the numbers 3 to 7, we can use `seq(3,7)` or, because they are consecutive, `3:7`. View the data and determine what rows are associated with the high fat or `hf` diet. Then use the `mean` function to compute the average weight of these mice.

6. One of the functions we will be using often is `sample`. Read the help file for `sample` using `?sample`. Now take a random sample of size 1 from the numbers 13 to 24 and report back the weight of the mouse represented by that row. Make sure to type `set.seed(1)` to ensure that everybody gets the same answer.

1.7   Brief Introduction to `dplyr`

The learning curve for R syntax is slow. One of the more difficult aspects that requires some getting used to is subsetting data tables. The `dplyr` package brings these tasks closer to English and we are therefore going to introduce two simple functions: one is used to subset and the other to select columns.

Take a look at the dataset we read in:

```
filename <- "femaleMiceWeights.csv"
dat <- read.csv(filename)
head(dat) #In R Studio use View(dat)
```

```
##   Diet Bodyweight
## 1 chow      21.51
## 2 chow      28.14
## 3 chow      24.04
## 4 chow      23.45
## 5 chow      23.68
## 6 chow      19.79
```

There are two types of diets, which are denoted in the first column. If we want just the weights, we only need the second column. So if we want the weights for mice on the `chow` diet, we subset and filter like this:

```
library(dplyr)
chow <- filter(dat, Diet=="chow") #keep only the ones with chow diet
head(chow)
```

```
##   Diet Bodyweight
## 1 chow      21.51
## 2 chow      28.14
## 3 chow      24.04
```

```
## 4 chow     23.45
## 5 chow     23.68
## 6 chow     19.79
```

And now we can select only the column with the values:

```
chowVals <- select(chow,Bodyweight)
head(chowVals)
```

```
##    Bodyweight
## 1      21.51
## 2      28.14
## 3      24.04
## 4      23.45
## 5      23.68
## 6      19.79
```

A nice feature of the **dplyr** package is that you can perform consecutive tasks by using what is called a "pipe". In **dplyr** we use %>% to denote a pipe. This symbol tells the program to first do one thing and then do something else to the result of the first. Hence, we can perform several data manipulations in one line. For example:

```
chowVals <- filter(dat, Diet=="chow") %>% select(Bodyweight)
```

In the second task, we no longer have to specify the object we are editing since it is whatever comes from the previous call.

Also, note that if **dplyr** receives a `data.frame` it will return a `data.frame`.

```
class(dat)
```

```
## [1] "data.frame"
```

```
class(chowVals)
```

```
## [1] "data.frame"
```

For pedagogical reasons, we will often want the final result to be a simple `numeric` vector. To obtain such a vector with **dplyr**, we can apply the **unlist** function which turns lists, such as `data.frames`, into `numeric` vectors:

```
chowVals <- filter(dat, Diet=="chow") %>% select(Bodyweight) %>% unlist
class( chowVals )
```

```
## [1] "numeric"
```

To do this in R without **dplyr** the code is the following:

```
chowVals <- dat[ dat$Diet=="chow", colnames(dat)=="Bodyweight"]
```

1.8   Exercises

For these exercises, we will use a new dataset related to mammalian sleep. This data is described here[16]. Download the CSV file from this[17] location. You can also download the data using the `downloader` package:

```
library(downloader)
dir <- "https://raw.githubusercontent.com/genomicsclass/dagdata/master/inst/extdata/"
filename <- "msleep_ggplot2.csv"
url <- paste0(dir, filename)
if (!file.exists(filename)) download(url,filename)
```

We are going to read in this data, then test your knowledge of they key `dplyr` functions `select` and `filter`. We are also going to review two different *classes*: data frames and vectors.

1. Read in the `msleep_ggplot2.csv` file with the function `read.csv` and use the function `class` to determine what type of object is returned.

2. Now use the `filter` function to select only the primates. How many animals in the table are primates? Hint: the `nrow` function gives you the number of rows of a data frame or matrix.

3. What is the class of the object you obtain after subsetting the table to only include primates?

4. Now use the `select` function to extract the sleep (total) for the primates. What class is this object? Hint: use `%>%` to pipe the results of the `filter` function to `select`.

5. Now we want to calculate the average amount of sleep for primates (the average of the numbers computed above). One challenge is that the `mean` function requires a vector so, if we simply apply it to the output above, we get an error. Look at the help file for `unlist` and use it to compute the desired average.

6. For the last exercise, we could also use the dplyr `summarize` function. We have not introduced this function, but you can read the help file and repeat exercise 5, this time using just `filter` and `summarize` to get the answer.

1.9   Mathematical Notation

This book focuses on teaching statistical concepts and data analysis programming skills. We avoid mathematical notation as much as possible, but we do use it. We do not want readers to be intimidated by the notation though. Mathematics is actually the easier part of learning statistics. Unfortunately, many text books use mathematical notation in what we believe to be an over-complicated way. For this reason, we do try to keep the notation as simple as possible. However, we do not want to water down the material, and some mathematical notation facilitates a deeper understanding of the concepts. Here we describe a few specific

---

[16]http://docs.ggplot2.org/0.9.3.1/msleep.html
[17]https://raw.githubusercontent.com/genomicsclass/dagdata/master/inst/extdata/msleep_ggplot2.csv

symbols that we use often. If they appear intimidating to you, please take some time to read this section carefully as they are actually simpler than they seem. Because by now you should be somewhat familiar with R, we will make the connection between mathematical notation and R code.

Indexing Those of us dealing with data almost always have a series of numbers. To describe the concepts in an abstract way, we use indexing. For example 5 numbers:

```
x <- 1:5
```

can be generally represented like this x_1, x_2, x_3, x_4, x_5. We use dots to simplify this $x_1, \ldots, x_5$ and indexing to simplify even more $x_i, i = 1, \ldots, 5$. If we want to describe a procedure for a list of any size n, we write $x_i, i = 1, \ldots, n$.

We sometimes have two indexes. For example, we may have several measurements (blood pressure, weight, height, age, cholesterol level) for 100 individuals. We can then use double indexes: $x_{i,j}, i = 1, \ldots, 100, j = 1, \ldots, 5$.

Summation A very common operation in data analysis is to sum several numbers. This comes up, for example, when we compute averages and standard deviations. If we have many numbers, there is a mathematical notation that makes it quite easy to express the following:

```
n <- 1000
x <- 1:n
S <- sum(x)
```

and it is the $\sum$ notation (capital S in Greek):

$$S = \sum_{i=1}^{n} x_i$$

Note that we make use of indexing as well. We will see that what is included inside the summation can become quite complicated. However, the summation part should not confuse you as it is a simple operation.

Greek letters We would prefer to avoid Greek letters, but they are ubiquitous in the statistical literature so we want you to become used to them. They are mainly used to distinguish the unknown from the observed. Suppose we want to find out the average height of a population and we take a sample of 1,000 people to estimate this. The unknown average we want to estimate is often denoted with μ, the Greek letter for m (m is for mean). The standard deviation is often denoted with σ, the Greek letter for s. Measurement error or other unexplained random variability is typically denoted with ε, the Greek letter for e. Effect sizes, for example the effect of a diet on weight, are typically denoted with β. We may use other Greek letters but those are the most commonly used.

You should get used to these four Greek letters as you will be seeing them often: μ, σ, β and ε.

Note that indexing is sometimes used in conjunction with Greek letters to denote different groups. For example, if we have one set of numbers denoted with x and another with y we may use μ_x and μ_y to denote their averages.

Infinity In the text we often talk about *asymptotic* results. Typically, this refers to an approximation that gets better and better as the number of data points we consider gets larger and larger, with perfect approximations occurring when the number of data points is ∞. In practice, there is no such thing as ∞, but it is a convenient concept to understand. One way to think about asymptotic results is as results that become better and better as some number increases and we can pick a number so that a computer can't tell the difference between the approximation and the real number. Here is a very simple example that approximates 1/3 with decimals:

```
onethird <- function(n) sum( 3/10^c(1:n))
1/3 - onethird(4)
```

```
## [1] 3.333333e-05
```

```
1/3 - onethird(10)
```

```
## [1] 3.333334e-11
```

```
1/3 - onethird(16)
```

```
## [1] 0
```

In the example above, 16 is practically ∞.

Integrals We only use these a couple of times so you can skip this section if you prefer. However, integrals are actually much simpler to understand than perhaps you realize.

For certain statistical operations, we need to figure out areas under the curve. For example, for a function $f(x)$ ...

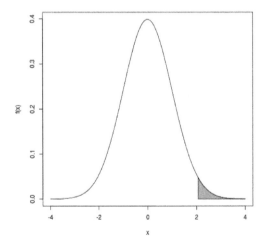

FIGURE 1.2
Integral of a function.

... we need to know what proportion of the total area under the curve is grey.

The grey area can be thought of as many small grey bars stacked next to each other. The area is then just the sum of the areas of these little bars. The problem is that we can't do this for every number between 2 and 4 because there are an infinite number. The integral is the mathematical solution to this problem. In this case, the total area is 1 so the answer to what proportion is grey is the following integral:

$$\int_2^4 f(x)\,dx$$

Because we constructed this example, we know that the grey area is 2.27% of the total. Note that this is very well approximated by an actual sum of little bars:

```
width <- 0.01
x <- seq(2,4,width)
areaofbars <-  f(x)*width
sum( areaofbars )
```

```
## [1] 0.02298998
```

The smaller we make `width`, the closer the sum gets to the integral, which is equal to the area.

2

Inference

2.1 Introduction

This chapter introduces the statistical concepts necessary to understand p-values and confidence intervals. These terms are ubiquitous in the life science literature. Let's use this paper[1] as an example.

Note that the abstract has this statement:

"Body weight was higher in mice fed the high-fat diet already after the first week, due to higher dietary intake in combination with lower metabolic efficiency."

To support this claim they provide the following in the results section:

"Already during the first week after introduction of high-fat diet, body weight increased significantly more in the high-fat diet-fed mice (+ 1.6 ± 0.1 g) than in the normal diet-fed mice (+ 0.2 ± 0.1 g; P < 0.001)."

What does P < 0.001 mean? What are the ± included? We will learn what this means and learn to compute these values in R. The first step is to understand random variables. To do this, we will use data from a mouse database (provided by Karen Svenson via Gary Churchill and Dan Gatti and partially funded by P50 GM070683). We will import the data into R and explain random variables and null distributions using R programming.

If you already downloaded the **femaleMiceWeights** file into your working directory, you can read it into R with just one line:

```
dat <- read.csv("femaleMiceWeights.csv")
```

Remember that a quick way to read the data, without downloading it is by using the url:

```
dir <- "https://raw.githubusercontent.com/genomicsclass/dagdata/master/inst/extdata/"
filename <- "femaleMiceWeights.csv"
url <- paste0(dir, filename)
dat <- read.csv(url)
```

Our first look at data We are interested in determining if following a given diet makes mice heavier after several weeks. This data was produced by ordering 24 mice from The Jackson Lab and randomly assigning either chow or high fat (hf) diet. After several weeks, the scientists weighed each mouse and obtained this data (**head** just shows us the first 6 rows):

```
head(dat)
```

---

[1] http://diabetes.diabetesjournals.org/content/53/suppl_3/S215.full

```
##    Diet Bodyweight
## 1 chow      21.51
## 2 chow      28.14
## 3 chow      24.04
## 4 chow      23.45
## 5 chow      23.68
## 6 chow      19.79
```

In RStudio, you can view the entire dataset with:

```
View(dat)
```

So are the hf mice heavier? Mouse 24 at 20.73 grams is one of the lightest mice, while Mouse 21 at 34.02 grams is one of the heaviest. Both are on the hf diet. Just from looking at the data, we see there is *variability*. Claims such as the one above usually refer to the averages. So let's look at the average of each group:

```
library(dplyr)
control <- filter(dat,Diet=="chow") %>% select(Bodyweight) %>% unlist
treatment <- filter(dat,Diet=="hf") %>% select(Bodyweight) %>% unlist
print( mean(treatment) )
```

```
## [1] 26.83417
```

```
print( mean(control) )
```

```
## [1] 23.81333
```

```
obsdiff <- mean(treatment) - mean(control)
print(obsdiff)
```

```
## [1] 3.020833
```

So the hf diet mice are about 10% heavier. Are we done? Why do we need p-values and confidence intervals? The reason is that these averages are random variables. They can take many values.

If we repeat the experiment, we obtain 24 new mice from The Jackson Laboratory and, after randomly assigning them to each diet, we get a different mean. Every time we repeat this experiment, we get a different value. We call this type of quantity a *random variable*.

2.2 Random Variables

Let's explore random variables further. Imagine that we actually have the weight of all control female mice and can upload them to R. In Statistics, we refer to this as *the population*. These are all the control mice available from which we sampled 24. Note that in practice we do not have access to the population. We have a special dataset that we are using here to illustrate concepts.

The first step is to download the data from here[2] into your working directory and then read it into R:

---

[2]https://raw.githubusercontent.com/genomicsclass/dagdata/master/inst/extdata/femaleControlsPopulation.csv

```
population <- read.csv("femaleControlsPopulation.csv")
##use unlist to turn it into a numeric vector
population <- unlist(population)
```

Now let's sample 12 mice three times and see how the average changes.

```
control <- sample(population,12)
mean(control)
```

```
## [1] 24.11333
```

```
control <- sample(population,12)
mean(control)
```

```
## [1] 24.40667
```

```
control <- sample(population,12)
mean(control)
```

```
## [1] 23.84
```

Note how the average varies. We can continue to do this repeatedly and start learning something about the distribution of this random variable.

2.3   The Null Hypothesis

Now let's go back to our average difference of `obsdiff`. As scientists we need to be skeptics. How do we know that this `obsdiff` is due to the diet? What happens if we give all 24 mice the same diet? Will we see a difference this big? Statisticians refer to this scenario as the *null hypothesis*. The name "null" is used to remind us that we are acting as skeptics: we give credence to the possibility that there is no difference.

Because we have access to the population, we can actually observe as many values as we want of the difference of the averages when the diet has no effect. We can do this by randomly sampling 24 control mice, giving them the same diet, and then recording the difference in mean between two randomly split groups of 12 and 12. Here is this process written in R code:

```
##12 control mice
control <- sample(population,12)
##another 12 control mice that we act as if they were not
treatment <- sample(population,12)
print(mean(treatment) - mean(control))
```

```
## [1] 0.5575
```

Now let's do it 10,000 times. We will use a "for-loop", an operation that lets us automate this (a simpler approach that, we will learn later, is to use `replicate`).

```
n <- 10000
null <- vector("numeric",n)
for (i in 1:n) {
```

```
  control <- sample(population,12)
  treatment <- sample(population,12)
  null[i] <- mean(treatment) - mean(control)
}
```

The values in `null` form what we call the *null distribution*. We will define this more formally below.

So what percent of the 10,000 are bigger than `obsdiff`?

```
mean(null >= obsdiff)
```

```
## [1] 0.0138
```

Only a small percent of the 10,000 simulations. As skeptics what do we conclude? When there is no diet effect, we see a difference as big as the one we observed only 1.5% of the time. This is what is known as a p-value, which we will define more formally later in the book.

2.4   Distributions

We have explained what we mean by *null* in the context of null hypothesis, but what exactly is a distribution? The simplest way to think of a *distribution* is as a compact description of many numbers. For example, suppose you have measured the heights of all men in a population. Imagine you need to describe these numbers to someone that has no idea what these heights are, such as an alien that has never visited Earth. Suppose all these heights are contained in the following dataset:

```
data(father.son,package="UsingR")
x <- father.son$fheight
```

One approach to summarizing these numbers is to simply list them all out for the alien to see. Here are 10 randomly selected heights of 1,078:

```
round(sample(x,10),1)
```

```
##  [1] 67.4 64.9 62.9 69.2 72.3 69.3 65.9 65.2 69.8 69.1
```

Cumulative Distribution Function   Scanning through these numbers, we start to get a rough idea of what the entire list looks like, but it is certainly inefficient. We can quickly improve on this approach by defining and visualizing a *distribution*. To define a distribution we compute, for all possible values of a, the proportion of numbers in our list that are below a. We use the following notation:

$$F(a) \equiv \Pr(x \le a)$$

This is called the cumulative distribution function (CDF). When the CDF is derived from data, as opposed to theoretically, we also call it the empirical CDF (ECDF). The ECDF for the height data looks like this:

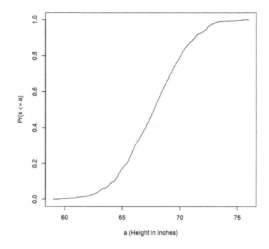

FIGURE 2.1
Empirical cummulative distribution function for height.

Histograms Although the empirical CDF concept is widely discussed in statistics text-books, the plot is actually not very popular in practice. The reason is that histograms give us the same information and are easier to interpret. Histograms show us the proportion of values in intervals:

$$\Pr(a \leq x \leq b) = F(b) - F(a)$$

Plotting these heights as bars is what we call a *histogram*. It is a more useful plot because we are usually more interested in intervals, such and such percent are between 70 inches and 71 inches, etc., rather than the percent less than a particular height. It is also easier to distinguish different types (families) of distributions by looking at histograms. Here is a histogram of heights:

```
hist(x)
```

We can specify the bins and add better labels in the following way:

```
bins <- seq(smallest, largest)
hist(x,breaks=bins,xlab="Height (in inches)",main="Adult men heights")
```

Showing this plot to the alien is much more informative than showing numbers. With this simple plot, we can approximate the number of individuals in any given interval. For example, there are about 70 individuals over six feet (72 inches) tall.

2.5 Probability Distribution

Summarizing lists of numbers is one powerful use of distribution. An even more important use is describing the possible outcomes of a random variable. Unlike a fixed list of numbers,

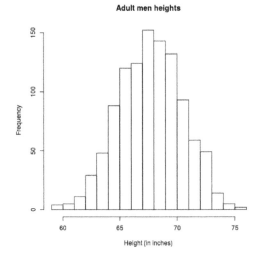

FIGURE 2.2
Histogram for heights.

we don't actually observe all possible outcomes of random variables, so instead of describing proportions, we describe probabilities. For instance, if we pick a random height from our list, then the probability of it falling between a and b is denoted with:

$$\Pr(a \leq X \leq b) = F(b) - F(a)$$

Note that the X is now capitalized to distinguish it as a random variable and that the equation above defines the probability distribution of the random variable. Knowing this distribution is incredibly useful in science. For example, in the case above, if we know the distribution of the difference in mean of mouse weights when the null hypothesis is true, referred to as the *null distribution*, we can compute the probability of observing a value as large as we did, referred to as a *p-value*. In a previous section we ran what is called a *Monte Carlo* simulation (we will provide more details on Monte Carlo simulation in a later section) and we obtained 10,000 outcomes of the random variable under the null hypothesis. Let's repeat the loop above, but this time let's add a point to the figure every time we re-run the experiment. If you run this code, you can see the null distribution forming as the observed values stack on top of each other.

```
n <- 100
library(rafalib)
nullplot(-5,5,1,30, xlab="Observed differences (grams)", ylab="Frequency")
totals <- vector("numeric",11)
for (i in 1:n) {
  control <- sample(population,12)
  treatment <- sample(population,12)
  nulldiff <- mean(treatment) - mean(control)
  j <- pmax(pmin(round(nulldiff)+6,11),1)
  totals[j] <- totals[j]+1
  text(j-6,totals[j],pch=15,round(nulldiff,1))
  ##if(i < 15) Sys.sleep(1) ##You can add this line to see values appear slowly
  }
```

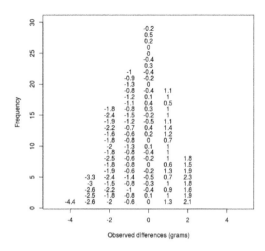

FIGURE 2.3
Illustration of the null distribution.

The figure above amounts to a histogram. From a histogram of the `null` vector we calculated earlier, we can see that values as large as `obsdiff` are relatively rare:

```
hist(null, freq=TRUE)
abline(v=obsdiff, col="red", lwd=2)
```

An important point to keep in mind here is that while we defined $\Pr(a)$ by counting cases, we will learn that, in some circumstances, mathematics gives us formulas for $\Pr(a)$ that save us the trouble of computing them as we did here. One example of this powerful approach uses the normal distribution approximation.

2.6   Normal Distribution

The probability distribution we see above approximates one that is very common in nature: the bell curve, also known as the normal distribution or Gaussian distribution. When the histogram of a list of numbers approximates the normal distribution, we can use a convenient mathematical formula to approximate the proportion of values or outcomes in any given interval:

$$\Pr(a < x < b) = \int_a^b \frac{1}{\sqrt{2\pi\sigma^2}} \exp\left(\frac{-(x-\mu)^2}{2\sigma^2}\right) dx$$

While the formula may look intimidating, don't worry, you will never actually have to type it out, as it is stored in a more convenient form (as `pnorm` in R which sets a to $-\infty$, and takes b as an argument).

Here μ and σ are referred to as the mean and the standard deviation of the population (we explain these in more detail in another section). If this *normal approximation* holds for our list, then the population mean and variance of our list can be used in the formula above. An example of this would be when we noted above that only 1.5% of values on the null

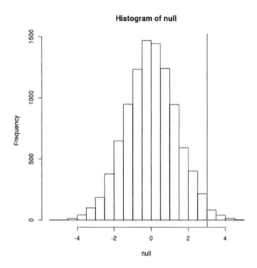

FIGURE 2.4
Null distribution with observed difference marked with vertical red line.

distribution were above `obsdiff`. We can compute the proportion of values below a value `x` with `pnorm(x,mu,sigma)` without knowing all the values. The normal approximation works very well here:

```
1 - pnorm(obsdiff,mean(null),sd(null))
```

```
## [1] 0.01391929
```

Later, we will learn that there is a mathematical explanation for this. A very useful characteristic of this approximation is that one only needs to know μ and σ to describe the entire distribution. From this, we can compute the proportion of values in any interval.

Summary So computing a p-value for the difference in diet for the mice was pretty easy, right? But why are we not done? To make the calculation, we did the equivalent of buying all the mice available from The Jackson Laboratory and performing our experiment repeatedly to define the null distribution. Yet this is not something we can do in practice. Statistical Inference is the mathematical theory that permits you to approximate this with only the data from your sample, i.e. the original 24 mice. We will focus on this in the following sections.

Setting the random seed Before we continue, we briefly explain the following important line of code:

```
set.seed(1)
```

Throughout this book, we use random number generators. This implies that many of the results presented can actually change by chance, including the correct answer to problems. One way to ensure that results do not change is by setting R's random number generation seed. For more on the topic please read the help file:

```
?set.seed
```

2.7 Exercises

For these exercises, we will be using the following dataset:

```
dir <- "https://raw.githubusercontent.com/genomicsclass/dagdata/master/inst/extdata/"
filename <- "femaleControlsPopulation.csv"
url <- paste0(dir, filename)
x <- unlist(read.csv(url))
```

Here x represents the weights for the entire population.

1. What is the average of these weights?

2. After setting the seed at 1, set.seed(1) take a random sample of size 5. What is the absolute value (use abs) of the difference between the average of the sample and the average of all the values?

3. After setting the seed at 5, set.seed(5) take a random sample of size 5. What is the absolute value of the difference between the average of the sample and the average of all the values?

4. Why are the answers from 2 and 3 different?

   (a) Because we made a coding mistake.
   (b) Because the average of the x is random.
   (c) Because the average of the samples is a random variable.
   (d) All of the above.

5. Set the seed at 1, then using a for-loop take a random sample of 5 mice 1,000 times. Save these averages. What percent of these 1,000 averages are more than 1 gram away from the average of x ?

6. We are now going to increase the number of times we redo the sample from 1,000 to 10,000. Set the seed at 1, then using a for-loop take a random sample of 5 mice 10,000 times. Save these averages. What percent of these 10,000 averages are more than 1 gram away from the average of x ?

7. Note that the answers to 5 and 6 barely changed. This is expected. The way we think about the random value distributions is as the distribution of the list of values obtained if we repeated the experiment an infinite number of times. On a computer, we can't perform an infinite number of iterations so instead, for our examples, we consider 1,000 to be large enough, thus 10,000 is as well. Now if instead we change the sample size, then we change the random variable and thus its distribution.

   Set the seed at 1, then using a for-loop take a random sample of 50 mice 1,000 times. Save these averages. What percent of these 1,000 averages are more than 1 gram away from the average of x ?

8. Use a histogram to "look" at the distribution of averages we get with a sample size of 5 and a sample size of 50. How would you say they differ?

   (a) They are actually the same.
   (b) They both look roughly normal, but with a sample size of 50 the spread is smaller.
   (c) They both look roughly normal, but with a sample size of 50 the spread is larger.

(d) The second distribution does not look normal at all.

9. For the last set of averages, the ones obtained from a sample size of 50, what percent are between 23 and 25?

10. Now ask the same question of a normal distribution with average 23.9 and standard deviation 0.43.

The answers to 9 and 10 were very similar. This is because we can approximate the distribution of the sample average with a normal distribution. We will learn more about the reason for this next.

2.8   Populations, Samples and Estimates

Now that we have introduced the idea of a random variable, a null distribution, and a p-value, we are ready to describe the mathematical theory that permits us to compute p-values in practice. We will also learn about confidence intervals and power calculations.

Population parameters A first step in statistical inference is to understand what population you are interested in. In the mouse weight example, we have two populations: female mice on control diets and female mice on high fat diets, with weight being the outcome of interest. We consider this population to be fixed, and the randomness comes from the sampling. One reason we have been using this dataset as an example is because we happen to have the weights of all the mice of this type. We download this[3] file to our working directory and read in to R:

```
dat <- read.csv("mice_pheno.csv")
```

We can then access the population values and determine, for example, how many we have. Here we compute the size of the control population:

```
library(dplyr)
controlPopulation <- filter(dat,Sex == "F" & Diet == "chow") %>%
  select(Bodyweight) %>% unlist
length(controlPopulation)
```

```
## [1] 225
```

We usually denote these values as $x_1, \ldots, x_m$. In this case, m is the number computed above. We can do the same for the high fat diet population:

```
hfPopulation <- filter(dat,Sex == "F" & Diet == "hf") %>%
  select(Bodyweight) %>% unlist
length(hfPopulation)
```

```
## [1] 200
```

and denote with $y_1, \ldots, y_n$.

We can then define summaries of interest for these populations, such as the mean and variance.

---

[3]https://raw.githubusercontent.com/genomicsclass/dagdata/master/inst/extdata/mice_pheno.csv

The mean:

$$\mu_X = \frac{1}{m}\sum_{i=1}^{m} x_i \text{ and } \mu_Y = \frac{1}{n}\sum_{i=1}^{n} y_i$$

The variance:

$$\sigma_X^2 = \frac{1}{m}\sum_{i=1}^{m}(x_i - \mu_X)^2 \text{ and } \sigma_Y^2 = \frac{1}{n}\sum_{i=1}^{n}(y_i - \mu_Y)^2$$

with the standard deviation being the square root of the variance. We refer to such quantities that can be obtained from the population as *population parameters*. The question we started out asking can now be written mathematically: is $\mu_Y - \mu_X = 0$?

Although in our illustration we have all the values and can check if this is true, in practice we do not. For example, in practice it would be prohibitively expensive to buy all the mice in a population. Here we learn how taking a *sample* permits us to answer our questions. This is the essence of statistical inference.

Sample estimates In the previous chapter, we obtained samples of 12 mice from each population. We represent data from samples with capital letters to indicate that they are random. This is common practice in statistics, although it is not always followed. So the samples are $X_1, \ldots, X_M$ and $Y_1, \ldots, Y_N$ and, in this case, $N = M = 12$. In contrast and as we saw above, when we list out the values of the population, which are set and not random, we use lower-case letters.

Since we want to know if $\mu_Y - \mu_X$ is 0, we consider the sample version: $\bar{Y} - \bar{X}$ with:

$$\bar{X} = \frac{1}{M}\sum_{i=1}^{M} X_i \text{ and } \bar{Y} = \frac{1}{N}\sum_{i=1}^{N} Y_i.$$

Note that this difference of averages is also a random variable. Previously, we learned about the behavior of random variables with an exercise that involved repeatedly sampling from the original distribution. Of course, this is not an exercise that we can execute in practice. In this particular case it would involve buying 24 mice over and over again. Here we described the mathematical theory that mathematically relates $\bar{X}$ to μ_X and $\bar{Y}$ to μ_Y, that will in turn help us understand the relationship between $\bar{Y} - \bar{X}$ and $\mu_Y - \mu_X$. Specifically, we will describe how the Central Limit Theorem permits us to use an approximation to answer this question, as well as motivate the widely used t-distribution.

2.9 Exercises

For these exercises, we will be using the following dataset:

```
dir <- "https://raw.githubusercontent.com/genomicsclass/dagdata/master/inst/extdata/"
filename <- "mice_pheno.csv"
url <- paste0(dir, filename)
dat <- read.csv(url)
```

We will remove the lines that contain missing values:

```
dat <- na.omit( dat )
```

1. Use `dplyr` to create a vector `x` with the body weight of all males on the control (`chow`) diet. What is this population's average?

2. Now use the `rafalib` package and use the `popsd` function to compute the population standard deviation.

3. Set the seed at 1. Take a random sample X of size 25 from `x`. What is the sample average?

4. Use `dplyr` to create a vector `y` with the body weight of all males on the high fat (`hf`) diet. What is this population's average?

5. Now use the `rafalib` package and use the `popsd` function to compute the population standard deviation.

6. Set the seed at 1. Take a random sample Y of size 25 from `y`. What is the sample average?

7. What is the difference in absolute value between $\bar{y} - \bar{x}$ and $\bar{Y} - \bar{X}$?

8. Repeat the above for females. Make sure to set the seed to 1 before each `sample` call. What is the difference in absolute value between $\bar{y} - \bar{x}$ and $\bar{Y} - \bar{X}$?

9. For the females, our sample estimates were closer to the population difference than with males. What is a possible explanation for this?

    (a) The population variance of the females is smaller than that of the males; thus, the sample variable has less variability.

    (b) Statistical estimates are more precise for females.

    (c) The sample size was larger for females.

    (d) The sample size was smaller for females.

2.10   Central Limit Theorem and t-distribution

Below we will discuss the Central Limit Theorem (CLT) and the t-distribution, both of which help us make important calculations related to probabilities. Both are frequently used in science to test statistical hypotheses. To use these, we have to make different assumptions from those for the CLT and the t-distribution. However, if the assumptions are true, then we are able to calculate the exact probabilities of events through the use of mathematical formula.

Central Limit Theorem   The CLT is one of the most frequently used mathematical results in science. It tells us that when the sample size is large, the average $\bar{Y}$ of a random sample follows a normal distribution centered at the population average μ_Y and with standard deviation equal to the population standard deviation σ_Y, divided by the square root of the sample size N. We refer to the standard deviation of the distribution of a random variable as the random variable's *standard error*.

Please note that if we subtract a constant from a random variable, the mean of the new random variable shifts by that constant. Mathematically, if X is a random variable with mean μ and a is a constant, the mean of $X - a$ is $\mu - a$. A similarly intuitive result holds for multiplication and the standard deviation (SD). If X is a random variable with mean μ and SD σ, and a is a constant, then the mean and SD of aX are $a\mu$ and $\mid a \mid \sigma$ respectively. To see how intuitive this is, imagine that we subtract 10 grams from each of the mice weights.

The average weight should also drop by that much. Similarly, if we change the units from grams to milligrams by multiplying by 1000, then the spread of the numbers becomes larger.

This implies that if we take many samples of size N, then the quantity:

$$\frac{\bar{Y} - \mu}{\sigma_Y/\sqrt{N}}$$

is approximated with a normal distribution centered at 0 and with standard deviation 1.

Now we are interested in the difference between two sample averages. Here again a mathematical result helps. If we have two random variables X and Y with means μ_X and μ_Y and variance σ_X and σ_Y respectively, then we have the following result: the mean of the sum $Y + X$ is the sum of the means $\mu_Y + \mu_X$. Using one of the facts we mentioned earlier, this implies that the mean of $Y - X = Y + aX$ with $a = -1$, which implies that the mean of $Y - X$ is $\mu_Y - \mu_X$. This is intuitive. However, the next result is perhaps not as intuitive. If X and Y are independent of each other, as they are in our mouse example, then the variance (SD squared) of $Y + X$ is the sum of the variances $\sigma_Y^2 + \sigma_X^2$. This implies that variance of the difference $Y - X$ is the variance of $Y + aX$ with $a = -1$ which is $\sigma_Y^2 + a^2\sigma_X^2 = \sigma_Y^2 + \sigma_X^2$. So the variance of the difference is also the sum of the variances. If this seems like a counterintuitive result, remember that if X and Y are independent of each other, the sign does not really matter. It can be considered random: if X is normal with certain variance, for example, so is $-X$. Finally, another useful result is that the sum of normal variables is again normal.

All this math is very helpful for the purposes of our study because we have two sample averages and are interested in the difference. Because both are normal, the difference is normal as well, and the variance (the standard deviation squared) is the sum of the two variances. Under the null hypothesis that there is no difference between the population averages, the difference between the sample averages $\bar{Y} - \bar{X}$, with $\bar{X}$ and $\bar{Y}$ the sample average for the two diets respectively, is approximated by a normal distribution centered at 0 (there is no difference) and with standard deviation $\sqrt{\sigma_X^2 + \sigma_Y^2}/\sqrt{N}$.

This suggests that this ratio:

$$\frac{\bar{Y} - \bar{X}}{\sqrt{\frac{\sigma_X^2}{M} + \frac{\sigma_Y^2}{N}}}$$

is approximated by a normal distribution centered at 0 and standard deviation 1. Using this approximation makes computing p-values simple because we know the proportion of the distribution under any value. For example, only 5% of these values are larger than 2 (in absolute value):

```
pnorm(-2) + (1 - pnorm(2))
```

```
## [1] 0.04550026
```

We don't need to buy more mice, 12 and 12 suffice.

However, we can't claim victory just yet because we don't know the population standard deviations: σ_X and σ_Y. These are unknown population parameters, but we can get around this by using the sample standard deviations, call them s_X and s_Y. These are defined as:

$$s_X^2 = \frac{1}{M-1}\sum_{i=1}^{M}(X_i - \bar{X})^2 \text{ and } s_Y^2 = \frac{1}{N-1}\sum_{i=1}^{N}(Y_i - \bar{Y})^2$$

Note that we are dividing by $M - 1$ and $N - 1$, instead of by M and N. There is a theoretical reason for doing this which we do not explain here. But to get an intuition, think of the case when you just have 2 numbers. The average distance to the mean is basically 1/2 the difference between the two numbers. So you really just have information from one number. This is somewhat of a minor point. The main point is that s_X and s_Y serve as estimates of σ_X and σ_Y

So we can redefine our ratio as

$$\sqrt{N}\frac{\bar{Y}-\bar{X}}{\sqrt{s_X^2+s_Y^2}}$$

if $M=N$ or in general,

$$\frac{\bar{Y}-\bar{X}}{\sqrt{\frac{s_X^2}{M}+\frac{s_Y^2}{N}}}$$

The CLT tells us that when M and N are large, this random variable is normally distributed with mean 0 and SD 1. Thus we can compute p-values using the function **pnorm**.

The t-distribution The CLT relies on large samples, what we refer to as *asymptotic results*. When the CLT does not apply, there is another option that does not rely on asymptotic results. When the original population from which a random variable, say Y, is sampled is normally distributed with mean 0, then we can calculate the distribution of:

$$\sqrt{N}\frac{\bar{Y}}{s_Y}$$

This is the ratio of two random variables so it is not necessarily normal. The fact that the denominator can be small by chance increases the probability of observing large values. William Sealy Gosset[4], an employee of the Guinness brewing company, deciphered the distribution of this random variable and published a paper under the pseudonym "Student". The distribution is therefore called Student's t-distribution. Later we will learn more about how this result is used.

Here we will use the mice phenotype data as an example. We start by creating two vectors, one for the control population and one for the high-fat diet population:

```
library(dplyr)
dat <- read.csv("mice_pheno.csv") #We downloaded this file in a previous section
controlPopulation <- filter(dat,Sex == "F" & Diet == "chow") %>%
  select(Bodyweight) %>% unlist
hfPopulation <- filter(dat,Sex == "F" & Diet == "hf") %>%
  select(Bodyweight) %>% unlist
```

It is important to keep in mind that what we are assuming to be normal here is the distribution of $y_1, y_2, \ldots, y_n$, not the random variable $\bar{Y}$. Although we can't do this in practice, in this illustrative example, we get to see this distribution for both controls and high fat diet mice:

```
library(rafalib)
mypar(1,2)
hist(hfPopulation)
hist(controlPopulation)
```

We can use *qq-plots* to confirm that the distributions are relatively close to being normally distributed. We will explore these plots in more depth in a later section, but the

---

[4]http://en.wikipedia.org/wiki/William_Sealy_Gosset

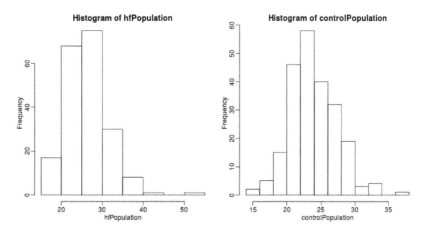

FIGURE 2.5
Histograms of all weights for both populations.

important thing to know is that it compares data (on the y-axis) against a theoretical distribution (on the x-axis). If the points fall on the identity line, then the data is close to the theoretical distribution.

```
mypar(1,2)
qqnorm(hfPopulation)
qqline(hfPopulation)
qqnorm(controlPopulation)
qqline(controlPopulation)
```

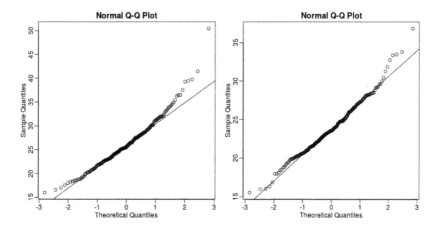

FIGURE 2.6
Quantile-quantile plots of all weights for both populations.

The larger the sample, the more forgiving the result is to the weakness of this approximation. In the next section, we will see that for this particular dataset the t-distribution works well even for sample sizes as small as 3.

2.11   Exercises

For these exercises, we will be using the following dataset:

```
dir <- "https://raw.githubusercontent.com/genomicsclass/dagdata/master/inst/extdata/"
filename <- "mice_pheno.csv"
url <- paste0(dir, filename)
##The data has missing values.
##We remove them with the na.omit function
dat <- na.omit( read.csv(url) )
```

1. If a list of numbers has a distribution that is well approximated by the normal distribution, what proportion of these numbers are within one standard deviation away from the list's average?

2. What proportion of these numbers are within two standard deviations away from the list's average?

3. What proportion of these numbers are within three standard deviations away from the list's average?

4. Define `y` to be the weights of males on the control diet. What proportion of the mice are within one standard deviation away from the average weight (remember to use `popsd` for the population `sd`)?

5. What proportion of these numbers are within two standard deviations away from the list's average?

6. What proportion of these numbers are within three standard deviations away from the list's average?

7. Note that the numbers for the normal distribution and our weights are relatively close. Also, notice that we are indirectly comparing quantiles of the normal distribution to quantiles of the mouse weight distribution. We can actually compare all quantiles using a qq-plot. Which of the following best describes the qq-plot comparing mouse weights to the normal distribution?

   (a) The points on the qq-plot fall exactly on the identity line.

   (b) The average of the mouse weights is not 0 and thus it can't follow a normal distribution.

   (c) The mouse weights are well approximated by the normal distribution, although the larger values (right tail) are larger than predicted by the normal. This is consistent with the differences seen between question 3 and 6.

   (d) These are not random variables and thus they can't follow a normal distribution.

8. Create the above qq-plot for the four populations: male/females on each of the two diets. What is the most likely explanation for the mouse weights being well approximated? What is the best explanation for all these being well approximated by the normal distribution?

   (a) The CLT tells us that sample averages are approximately normal.

   (b) This just happens to be how nature behaves. Perhaps the result of many biological factors averaging out.

   (c) Everything measured in nature follows a normal distribution.

   (d) Measurement error is normally distributed.

9. Here we are going to use the function `replicate` to learn about the distribution of random variables. All the above exercises relate to the normal distribution as an approximation of the distribution of a fixed list of numbers or a population. We have not yet discussed probability in these exercises. If the distribution of a list of numbers is approximately normal, then if we pick a number at random from this distribution, it will follow a normal distribution. However, it is important to remember that stating that some quantity has a distribution does not necessarily imply this quantity is random. Also, keep in mind that this is not related to the central limit theorem. The central limit applies to averages of random variables. Let's explore this concept.

We will now take a sample of size 25 from the population of males on the chow diet. The average of this sample is our random variable. We will use the `replicate` to observe 10,000 realizations of this random variable. Set the seed at 1, generate these 10,000 averages. Make a histogram and qq-plot of these 10,000 numbers against the normal distribution.

We can see that, as predicted by the CLT, the distribution of the random variable is very well approximated by the normal distribution.

```
y <- filter(dat, Sex=="M" & Diet=="chow") %>% select(Bodyweight) %>% unlist
avgs <- replicate(10000, mean( sample(y, 25)))
mypar(1,2)
hist(avgs)
qqnorm(avgs)
    qqline(avgs)
```

What is the average of the distribution of the sample average?

10. What is the standard deviation of the distribution of sample averages?

11. According to the CLT, the answer to exercise 9 should be the same as `mean(y)`. You should be able to confirm that these two numbers are very close. Which of the following does the CLT tell us should be close to your answer to exercise 10?

    (a) `popsd(y)`
    (b) `popsd(avgs)/sqrt(25)`
    (c) `sqrt(25) / popsd(y)`
    (d) `popsd(y)/sqrt(25)`

12. In practice we do not know σ (`popsd(y)`) which is why we can't use the CLT directly. This is because we see a sample and not the entire distribution. We also can't use `popsd(avgs)` because to construct averages, we have to take 10,000 samples and this is never practical. We usually just get one sample. Instead we have to estimate `popsd(y)`. As described, what we use is the sample standard deviation. Set the seed at 1, using the `replicate` function, create 10,000 samples of 25 and now, instead of the sample average, keep the standard deviation. Look at the distribution of the sample standard deviations. It is a random variable. The real population SD is about 4.5. What proportion of the sample SDs are below 3.5?

13. What the answer to question 12 reveals is that the denominator of the t-test is a random variable. By decreasing the sample size, you can see how this variability can increase. It therefore adds variability. The smaller the sample size, the more variability is added. The normal distribution stops providing a useful approximation. When the distribution of the population values is approximately

normal, as it is for the weights, the t-distribution provides a better approxima-tion. We will see this later on. Here we will look at the difference between the t-distribution and normal. Use the function `qt` and `qnorm` to get the quantiles of `x=seq(0.0001,0.9999,len=300)`. Do this for degrees of freedom 3, 10, 30, and 100. Which of the following is true?

(a) The t-distribution and normal distribution are always the same.
(b) The t-distribution has a higher average than the normal distribution.
(c) The t-distribution has larger tails up until 30 degrees of freedom, at which point it is practically the same as the normal distribution.
(d) The variance of the t-distribution grows as the degrees of freedom grow.

2.12   Central Limit Theorem in Practice

Let's use our data to see how well the central limit theorem approximates sample averages from our data. We will leverage our entire population dataset to compare the results we obtain by actually sampling from the distribution to what the CLT predicts.

```
dat <- read.csv("mice_pheno.csv") #file was previously downloaded
head(dat)
```

```
##   Sex Diet Bodyweight
## 1   F   hf      31.94
## 2   F   hf      32.48
## 3   F   hf      22.82
## 4   F   hf      19.92
## 5   F   hf      32.22
## 6   F   hf      27.50
```

Start by selecting only female mice since males and females have different weights. We will select three mice from each population.

```
library(dplyr)
controlPopulation <- filter(dat,Sex == "F" & Diet == "chow") %>%
  select(Bodyweight) %>% unlist
hfPopulation <- filter(dat,Sex == "F" & Diet == "hf") %>%
  select(Bodyweight) %>% unlist
```

We can compute the population parameters of interest using the mean function.

```
mu_hf <- mean(hfPopulation)
mu_control <- mean(controlPopulation)
print(mu_hf - mu_control)
```

```
## [1] 2.375517
```

We can compute the population standard deviations of, say, a vector x as well. However, we do not use the R function `sd` because this function actually does not compute the population standard deviation σ_x. Instead, `sd` assumes the main argument is a random sample, say X, and provides an estimate of σ_x, defined by s_X above. As shown in the equations above the actual final answer differs because one divides by the sample size and the other by the sample size minus one. We can see that with R code:

```
x <- controlPopulation
N <- length(x)
populationvar <- mean((x-mean(x))^2)
identical(var(x), populationvar)
```

```
## [1] FALSE
```

```
identical(var(x)*(N-1)/N, populationvar)
```

```
## [1] TRUE
```

So to be mathematically correct, we do not use `sd` or `var`. Instead, we use the `popvar` and `popsd` function in `rafalib`:

```
library(rafalib)
sd_hf <- popsd(hfPopulation)
sd_control <- popsd(controlPopulation)
```

Remember that in practice we do not get to compute these population parameters. These are values we never see. In general, we want to estimate them from samples.

```
N <- 12
hf <- sample(hfPopulation, 12)
control <- sample(controlPopulation, 12)
```

As we described, the CLT tells us that for large N, each of these is approximately normal with average population mean and standard error population variance divided by N. We mentioned that a rule of thumb is that N should be 30 or more. However, that is just a rule of thumb since the preciseness of the approximation depends on the population distribution. Here we can actually check the approximation and we do that for various values of N.

Now we use `sapply` and `replicate` instead of `for` loops, which makes for cleaner code (we do not have to pre-allocate a vector, R takes care of this for us):

```
Ns <- c(3,12,25,50)
B <- 10000 #number of simulations
res <- sapply(Ns,function(n) {
  replicate(B,mean(sample(hfPopulation,n))-mean(sample(controlPopulation,n)))
})
```

Now we can use qq-plots to see how well CLT approximations works for these. If in fact the normal distribution is a good approximation, the points should fall on a straight line when compared to normal quantiles. The more it deviates, the worse the approximation. In the title, we also show the average and SD of the observed distribution, which demonstrates how the SD decreases with $\sqrt{N}$ as predicted.

```
mypar(2,2)
for (i in seq(along=Ns)) {
  titleavg <- signif(mean(res[,i]),3)
  titlesd <- signif(popsd(res[,i]),3)
  title <- paste0("N=",Ns[i]," Avg=",titleavg," SD=",titlesd)
  qqnorm(res[,i],main=title)
  qqline(res[,i],col=2)
}
```

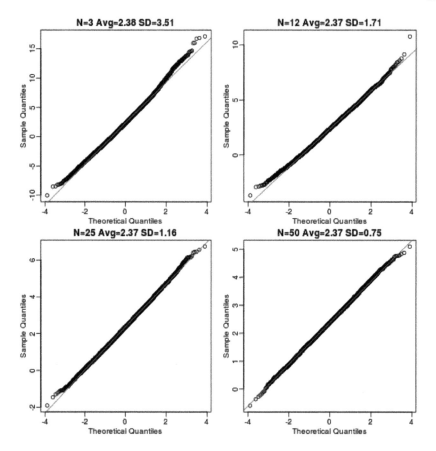

FIGURE 2.7
Quantile versus quantile plot of simulated differences versus theoretical normal distribution for four different sample sizes.

Here we see a pretty good fit even for 3. Why is this? Because the population itself is relatively close to normally distributed, the averages are close to normal as well (the sum of normals is also a normal). In practice, we actually calculate a ratio: we divide by the estimated standard deviation. Here is where the sample size starts to matter more.

```
Ns <- c(3,12,25,50)
B <- 10000 #number of simulations
##function to compute a t-stat
computetstat <- function(n) {
  y <- sample(hfPopulation,n)
  x <- sample(controlPopulation,n)
  (mean(y)-mean(x))/sqrt(var(y)/n+var(x)/n)
}
res <-  sapply(Ns,function(n) {
  replicate(B,computetstat(n))
})
mypar(2,2)
for (i in seq(along=Ns)) {
  qqnorm(res[,i],main=Ns[i])
```

```
    qqline(res[,i],col=2)
}
```

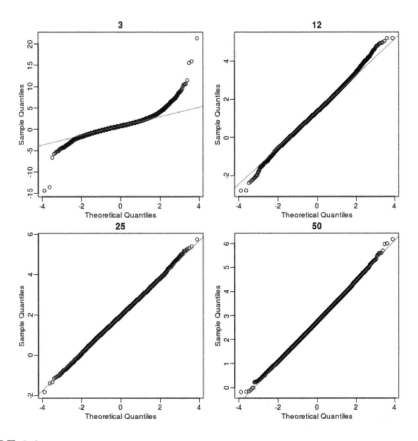

FIGURE 2.8
Quantile versus quantile plot of simulated ratios versus theoretical normal distribution for four different sample sizes.

So we see that for $N = 3$, the CLT does not provide a usable approximation. For $N = 12$, there is a slight deviation at the higher values, although the approximation appears useful. For 25 and 50, the approximation is spot on.

This simulation only proves that $N = 12$ is large enough in this case, not in general. As mentioned above, we will not be able to perform this simulation in most situations. We only use the simulation to illustrate the concepts behind the CLT and its limitations. In future sections, we will describe the approaches we actually use in practice.

2.13 Exercises

Exercises 3-13 use the mouse dataset we have previously downloaded:

```
dir <- "https://raw.githubusercontent.com/genomicsclass/dagdata/master/inst/extdata/"
filename <- "femaleMiceWeights.csv"
url <- paste0(dir, filename)
dat <- read.csv(url)
```

1. The CLT is a result from probability theory. Much of probability theory was originally inspired by gambling. This theory is still used in practice by casinos. For example, they can estimate how many people need to play slots for there to be a 99.9999% probability of earning enough money to cover expenses. Let's try a simple example related to gambling.

Suppose we are interested in the proportion of times we see a 6 when rolling `n=100` die. This is a random variable which we can simulate with `x=sample(1:6, n, replace=TRUE)` and the proportion we are interested in can be expressed as an average: `mean(x==6)`. Because the die rolls are independent, the CLT applies.

We want to roll `n` dice 10,000 times and keep these proportions. This random variable (proportion of 6s) has mean `p=1/6` and variance `p*(1-p)/n`. So according to CLT `z = (mean(x==6) - p) / sqrt(p*(1-p)/n)` should be normal with mean 0 and SD 1. Set the seed to 1, then use `replicate` to perform the simulation, and report what proportion of times `z` was larger than 2 in absolute value (CLT says it should be about 0.05).

2. For the last simulation you can make a qqplot to confirm the normal approximation. Now, the CLT is an *asymptotic* result, meaning it is closer and closer to being a perfect approximation as the sample size increases. In practice, however, we need to decide if it is appropriate for actual sample sizes. Is 10 enough? 15? 30?

In the example used in exercise 1, the original data is binary (either 6 or not). In this case, the success probability also affects the appropriateness of the CLT. With very low probabilities, we need larger sample sizes for the CLT to "kick in".

Run the simulation from exercise 1, but for different values of `p` and `n`. For which of the following is the normal approximation best?

- A) `p=0.5` and `n=5`

- B) `p=0.5` and `n=30`

- C) `p=0.01` and `n=30`

- D) `p=0.01` and `n=100`

3. As we have already seen, the CLT also applies to averages of quantitative data. A major difference with binary data, for which we know the variance is $p(1-p)$, is that with quantitative data we need to estimate the population standard deviation.

In several previous exercises we have illustrated statistical concepts with the unrealistic situation of having access to the entire population. In practice, we do *not* have access to entire populations. Instead, we obtain one random sample and need to reach conclusions analyzing that data. `dat` is an example of a typical simple dataset representing just one sample. We have 12 measurements for each of two populations:

```
X <- filter(dat, Diet=="chow") %>% select(Bodyweight) %>% unlist
Y <- filter(dat, Diet=="hf") %>% select(Bodyweight) %>% unlist
```

We think of X as a random sample from the population of all mice in the control diet and Y as a random sample from the population of all mice in the high fat diet.

Define the parameter μ_x as the average of the control population. We estimate this parameter with the sample average $\bar{X}$. What is the sample average?

4. We don't know μ_X, but want to use $\bar{X}$ to understand μ_X. Which of the following uses CLT to understand how well $\bar{X}$ approximates μ_X?

   (a) $\bar{X}$ follows a normal distribution with mean 0 and standard deviation 1.

   (b) μ_X follows a normal distribution with mean $\bar{X}$ and standard deviation $\frac{\sigma_x}{\sqrt{12}}$ where σ_x is the population standard deviation.

   (c) $\bar{X}$ follows a normal distribution with mean μ_X and standard deviation σ_x where σ_x is the population standard deviation.

   (d) $\bar{X}$ follows a normal distribution with mean μ_X and standard deviation $\frac{\sigma_x}{\sqrt{12}}$ where σ_x is the population standard deviation.

5. The result above tells us the distribution of the following random variable: $Z = \sqrt{12}\frac{\bar{X}-\mu_X}{\sigma_X}$. What does the CLT tell us is the mean of Z (you don't need code)?

6. The result of 4 and 5 tell us that we know the distribution of the difference between our estimate and what we want to estimate, but don't know. However, the equation involves the population standard deviation σ_X, which we don't know. Given what we discussed, what is your estimate of σ_x?

7. Use the CLT to approximate the probability that our estimate $\bar{X}$ is off by more than 2 grams from μ_X.

8. Now we introduce the concept of a null hypothesis. We don't know μ_x nor μ_y. We want to quantify what the data say about the possibility that the diet has no effect: $\mu_x = \mu_y$. If we use CLT, then we approximate the distribution of $\bar{X}$ as normal with mean μ_X and standard deviation $\sigma_X/\sqrt{M}$ and the distribution of $\bar{Y}$ as normal with mean μ_y and standard deviation $\sigma_y\sqrt{N}$, with M and N as the sample sizes for X and Y respectively, in this case 12. This implies that the difference $\bar{Y} - \bar{X}$ has mean 0. We described that the standard deviation of this statistic (the standard error) is $\mathrm{SE}(\bar{X} - \bar{Y}) = \sqrt{\sigma_y^2/12 + \sigma_x^2/12}$ and that we estimate the population standard deviations σ_x and σ_y with the sample estimates. What is the estimate of $\mathrm{SE}(\bar{X} - \bar{Y}) = \sqrt{\sigma_y^2/12 + \sigma_x^2/12}$?

9. So now we can compute $\bar{Y} - \bar{X}$ as well as an estimate of this standard error and construct a t-statistic. What is this t-statistic?

10. If we apply the CLT, what is the distribution of this t-statistic?

   (a) Normal with mean 0 and standard deviation 1.

   (b) t-distributed with 22 degrees of freedom.

   (c) Normal with mean 0 and standard deviation $\sqrt{\sigma_y^2/12 + \sigma_x^2/12}$.

   (d) t-distributed with 12 degrees of freedom.

11. Now we are ready to compute a p-value using the CLT. What is the probability of observing a quantity as large as what we computed in 10, when the null distribution is true?

12. CLT provides an approximation for cases in which the sample size is large. In practice, we can't check the assumption because we only get to see 1 outcome (which you computed above). As a result, if this approximation is off, so is our p-value. As described earlier, there is another approach that does not require a large sample size, but rather that the distribution of the population is approximately normal. We don't get to see this distribution so it is again an assumption, although we can look at the distribution of the sample with `qqnorm(X)` and `qqnorm(Y)`. If we are willing to assume this, then it follows that the t-statistic follows t-distribution. What is the p-value under the t-distribution approximation? Hint: use the `t.test` function.

13. With the CLT distribution, we obtained a p-value smaller than 0.05 and with the t-distribution, one that is larger. They can't both be right. What best describes the difference?

    (a) A sample size of 12 is not large enough, so we have to use the t-distribution approximation.

    (b) These are two different assumptions. The t-distribution accounts for the variability introduced by the estimation of the standard error and thus, under the null, large values are more probable under the null distribution.

    (c) The population data is probably not normally distributed so the t-distribution approximation is wrong.

    (d) Neither assumption is useful. Both are wrong.

2.14 t-tests in Practice

Introduction We will now demonstrate how to obtain a p-value in practice. We begin by loading experimental data and walking you through the steps used to form a t-statistic and compute a p-value. We can perform this task with just a few lines of code (go to the end of this section to see them). However, to understand the concepts, we will construct a t-statistic from "scratch".

Read in and prepare data We start by reading in the data. A first important step is to identify which rows are associated with treatment and control, and to compute the difference in mean.

```
library(dplyr)
dat <- read.csv("femaleMiceWeights.csv") #previously downloaded

control <- filter(dat,Diet=="chow") %>% select(Bodyweight) %>% unlist
treatment <- filter(dat,Diet=="hf") %>% select(Bodyweight) %>% unlist

diff <- mean(treatment) - mean(control)
print(diff)
```

```
## [1] 3.020833
```

We are asked to report a p-value. What do we do? We learned that `diff`, referred to as the *observed effect size*, is a random variable. We also learned that under the null hypothesis, the mean of the distribution of `diff` is 0. What about the standard error? We also learned that the standard error of this random variable is the population standard deviation divided by the square root of the sample size:

$$SE(\bar{X}) = \sigma/\sqrt{N}$$

We use the sample standard deviation as an estimate of the population standard deviation. In R, we simply use the `sd` function and the SE is:

```
sd(control)/sqrt(length(control))
```

```
## [1] 0.8725323
```

This is the SE of the sample average, but we actually want the SE of `diff`. We saw how statistical theory tells us that the variance of the difference of two random variables is the sum of its variances, so we compute the variance and take the square root:

```
se <- sqrt(
  var(treatment)/length(treatment) +
  var(control)/length(control)
  )
```

Statistical theory tells us that if we divide a random variable by its SE, we get a new random variable with an SE of 1.

```
tstat <- diff/se
```

This ratio is what we call the t-statistic. It's the ratio of two random variables and thus a random variable. Once we know the distribution of this random variable, we can then easily compute a p-value.

As explained in the previous section, the CLT tells us that for large sample sizes, both sample averages `mean(treatment)` and `mean(control)` are normal. Statistical theory tells us that the difference of two normally distributed random variables is again normal, so CLT tells us that `tstat` is approximately normal with mean 0 (the null hypothesis) and SD 1 (we divided by its SE).

So now to calculate a p-value all we need to do is ask: how often does a normally distributed random variable exceed `diff`? R has a built-in function, `pnorm`, to answer this specific question. `pnorm(a)` returns the probability that a random variable following the standard normal distribution falls below `a`. To obtain the probability that it is larger than `a`, we simply use `1-pnorm(a)`. We want to know the probability of seeing something as extreme as `diff`: either smaller (more negative) than `-abs(diff)` or larger than `abs(diff)`. We call these two regions "tails" and calculate their size:

```
righttail <- 1 - pnorm(abs(tstat))
lefttail <- pnorm(-abs(tstat))
pval <- lefttail + righttail
print(pval)
```

```
## [1] 0.0398622
```

In this case, the p-value is smaller than 0.05 and using the conventional cutoff of 0.05, we would call the difference *statistically significant*.

Now there is a problem. CLT works for large samples, but is 12 large enough? A rule of thumb for CLT is that 30 is a large enough sample size (but this is just a rule of thumb). The p-value we computed is only a valid approximation if the assumptions hold, which do not seem to be the case here. However, there is another option other than using CLT.

2.15 The t-distribution in Practice

As described earlier, statistical theory offers another useful result. If the distribution of the population is normal, then we can work out the exact distribution of the t-statistic without the need for the CLT. This is a big "if" given that, with small samples, it is hard to check if the population is normal. But for something like weight, we suspect that the population distribution is likely well approximated by normal and that we can use this approximation. Furthermore, we can look at a qq-plot for the samples. This shows that the approximation is at least close:

```
library(rafalib)
mypar(1,2)

qqnorm(treatment)
qqline(treatment,col=2)

qqnorm(control)
qqline(control,col=2)
```

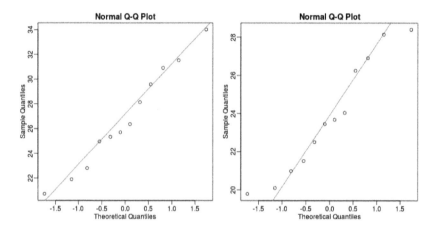

FIGURE 2.9
Quantile-quantile plots for sample against theoretical normal distribution.

If we use this approximation, then statistical theory tells us that the distribution of the random variable `tstat` follows a t-distribution. This is a much more complicated distribution than the normal. The t-distribution has a location parameter like the normal and another parameter called *degrees of freedom*. R has a nice function that actually computes everything for us.

```
t.test(treatment, control)

##
##   Welch Two Sample t-test
##
## data:  treatment and control
## t = 2.0552, df = 20.236, p-value = 0.053
## alternative hypothesis: true difference in means is not equal to 0
## 95 percent confidence interval:
##   -0.04296563  6.08463229
## sample estimates:
## mean of x mean of y
##   26.83417  23.81333
```

To see just the p-value, we can use the $ extractor:

```
result <- t.test(treatment,control)
result$p.value
```

```
## [1] 0.05299888
```

The p-value is slightly bigger now. This is to be expected because our CLT approximation considered the denominator of `tstat` practically fixed (with large samples it practically is), while the t-distribution approximation takes into account that the denominator (the standard error of the difference) is a random variable. The smaller the sample size, the more the denominator varies.

It may be confusing that one approximation gave us one p-value and another gave us another, because we expect there to be just one answer. However, this is not uncommon in data analysis. We used different assumptions, different approximations, and therefore we obtained different results.

Later, in the power calculation section, we will describe type I and type II errors. As a preview, we will point out that the test based on the CLT approximation is more likely to incorrectly reject the null hypothesis (a false positive), while the t-distribution is more likely to incorrectly accept the null hypothesis (false negative).

Running the t-test in practice Now that we have gone over the concepts, we can show the relatively simple code that one would use to actually compute a t-test:

```
library(dplyr)
dat <- read.csv("mice_pheno.csv")
control <- filter(dat,Diet=="chow") %>% select(Bodyweight)
treatment <- filter(dat,Diet=="hf") %>% select(Bodyweight)
t.test(treatment,control)
```

```
##
##  Welch Two Sample t-test
##
## data:  treatment and control
## t = 7.1932, df = 735.02, p-value = 1.563e-12
## alternative hypothesis: true difference in means is not equal to 0
## 95 percent confidence interval:
##   2.231533 3.906857
## sample estimates:
## mean of x mean of y
##   30.48201  27.41281
```

The arguments to `t.test` can be of type *data.frame* and thus we do not need to unlist them into numeric objects.

2.16   Confidence Intervals

We have described how to compute p-values which are ubiquitous in the life sciences. However, we do not recommend reporting p-values as the only statistical summary of your results. The reason is simple: statistical significance does not guarantee scientific significance.

With large enough sample sizes, one might detect a statistically significance difference in weight of, say, 1 microgram. But is this an important finding? Would we say a diet results in higher weight if the increase is less than a fraction of a percent? The problem with reporting only p-values is that you will not provide a very important piece of information: the effect size. Recall that the effect size is the observed difference. Sometimes the effect size is divided by the mean of the control group and so expressed as a percent increase.

A much more attractive alternative is to report confidence intervals. A confidence interval includes information about your estimated effect size and the uncertainty associated with this estimate. Here we use the mice data to illustrate the concept behind confidence intervals.

Confidence Interval for Population Mean Before we show how to construct a confidence interval for the difference between the two groups, we will show how to construct a confidence interval for the population mean of control female mice. Then we will return to the group difference after we've learned how to build confidence intervals in the simple case. We start by reading in the data and selecting the appropriate rows:

```
dat <- read.csv("mice_pheno.csv")
chowPopulation <- dat[dat$Sex=="F" & dat$Diet=="chow",3]
```

The population average μ_X is our parameter of interest here:

```
mu_chow <- mean(chowPopulation)
print(mu_chow)
```

```
## [1] 23.89338
```

We are interested in estimating this parameter. In practice, we do not get to see the entire population so, as we did for p-values, we demonstrate how we can use samples to do this. Let's start with a sample of size 30:

```
N <- 30
chow <- sample(chowPopulation,N)
print(mean(chow))
```

```
## [1] 23.351
```

We know this is a random variable, so the sample average will not be a perfect estimate. In fact, because in this illustrative example we know the value of the parameter, we can see that they are not exactly the same. A confidence interval is a statistical way of reporting our finding, the sample average, in a way that explicitly summarizes the variability of our random variable.

With a sample size of 30, we will use the CLT. The CLT tells us that $\bar{X}$ or `mean(chow)` follows a normal distribution with mean μ_X or `mean(chowPopulation)` and standard error approximately $s_X/\sqrt{N}$ or:

```
se <- sd(chow)/sqrt(N)
print(se)
```

```
## [1] 0.4781652
```

Defining the Interval A 95% confidence interval (we can use percentages other than 95%) is a random interval with a 95% probability of falling on the parameter we are estimating. Keep in mind that saying 95% of random intervals will fall on the true value (our definition above) is *not the same* as saying there is a 95% chance that the true value falls in our interval. To construct it, we note that the CLT tells us that $\sqrt{N}(\bar{X} - \mu_X)/s_X$ follows a normal distribution with mean 0 and SD 1. This implies that the probability of this event:

$$-2 \leq \sqrt{N}(\bar{X} - \mu_X)/s_X \leq 2$$

which written in R code is:

```
pnorm(2) - pnorm(-2)
```

```
## [1] 0.9544997
```

...is about 95% (to get closer use `qnorm(1-0.05/2)` instead of 2). Now do some basic algebra to clear out everything and leave μ_X alone in the middle and you get that the following event:

$$\bar{X} - 2s_X/\sqrt{N} \leq \mu_X \leq \bar{X} + 2s_X/\sqrt{N}$$

has a probability of 95%.

Be aware that it is the edges of the interval $\bar{X} \pm 2s_X/\sqrt{N}$, not μ_X, that are random. Again, the definition of the confidence interval is that 95% of *random intervals* will contain the true, fixed value μ_X. For a specific interval that has been calculated, the probability is either 0 or 1 that it contains the fixed population mean μ_X.

Let's demonstrate this logic through simulation. We can construct this interval with R relatively easily:

```
Q <- qnorm(1- 0.05/2)
interval <- c(mean(chow)-Q*se, mean(chow)+Q*se )
interval
```

```
## [1] 22.41381 24.28819
```

```
interval[1] < mu_chow & interval[2] > mu_chow
```

```
## [1] TRUE
```

which happens to cover μ_X or `mean(chowPopulation)`. However, we can take another sample and we might not be as lucky. In fact, the theory tells us that we will cover μ_X 95% of the time. Because we have access to the population data, we can confirm this by taking several new samples:

```
library(rafalib)
B <- 250
mypar()
plot(mean(chowPopulation)+c(-7,7),c(1,1),type="n",
     xlab="weight",ylab="interval",ylim=c(1,B))
abline(v=mean(chowPopulation))
for (i in 1:B) {
  chow <- sample(chowPopulation,N)
  se <- sd(chow)/sqrt(N)
  interval <- c(mean(chow)-Q*se, mean(chow)+Q*se)
  covered <-
```

```
      mean(chowPopulation) <= interval[2] & mean(chowPopulation) >= interval[1]
    color <- ifelse(covered,1,2)
    lines(interval, c(i,i),col=color)
}
```

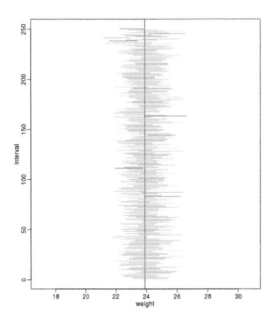

FIGURE 2.10
We show 250 random realizations of 95% confidence intervals. The color denotes if the
interval fell on the parameter or not.

You can run this repeatedly to see what happens. You will see that in about 5% of the
cases, we fail to cover μ_X.

Small Sample Size and the CLT For $N = 30$, the CLT works very well. However, if
$N = 5$, do these confidence intervals work as well? We used the CLT to create our intervals,
and with $N = 5$ it may not be as useful an approximation. We can confirm this with a
simulation:

```
mypar()
plot(mean(chowPopulation)+c(-7,7),c(1,1),type="n",
    xlab="weight",ylab="interval",ylim=c(1,B))
abline(v=mean(chowPopulation))
Q <- qnorm(1- 0.05/2)
N <- 5
for (i in 1:B) {
  chow <- sample(chowPopulation,N)
  se <- sd(chow)/sqrt(N)
  interval <- c(mean(chow)-Q*se, mean(chow)+Q*se)
  covered <- mean(chowPopulation) <= interval[2] & mean(chowPopulation) >= interval[1]
  color <- ifelse(covered,1,2)
  lines(interval, c(i,i),col=color)
}
```

Despite the intervals being larger (we are dividing by $\sqrt{5}$ instead of $\sqrt{30}$), we see many
more intervals not covering μ_X. This is because the CLT is incorrectly telling us that the
distribution of the `mean(chow)` is approximately normal with standard deviation 1 when, in

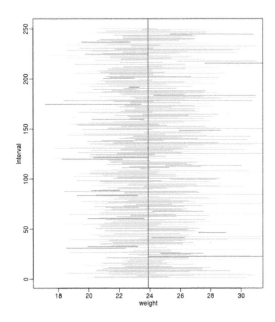

FIGURE 2.11

We show 250 random realizations of 95% confidence intervals, but now for a smaller sample size. The confidence interval is based on the CLT approximation. The color denotes if the interval fell on the parameter or not.

fact, it has a larger standard deviation and a fatter tail (the parts of the distribution going to $\pm\infty$). This mistake affects us in the calculation of Q, which assumes a normal distribution and uses **qnorm**. The t-distribution might be more appropriate. All we have to do is re-run the above, but change how we calculate Q to use **qt** instead of **qnorm**.

```
mypar()
plot(mean(chowPopulation) + c(-7,7), c(1,1), type="n",
     xlab="weight", ylab="interval", ylim=c(1,B))
abline(v=mean(chowPopulation))
##Q <- qnorm(1- 0.05/2) ##no longer normal so use:
Q <- qt(1- 0.05/2, df=4)
N <- 5
for (i in 1:B) {
  chow <- sample(chowPopulation, N)
  se <- sd(chow)/sqrt(N)
  interval <- c(mean(chow)-Q*se, mean(chow)+Q*se )
  covered <- mean(chowPopulation) <= interval[2] & mean(chowPopulation) >= interval[1]
  color <- ifelse(covered,1,2)
  lines(interval, c(i,i),col=color)
}
```

Now the intervals are made bigger. This is because the t-distribution has fatter tails and therefore:

```
qt(1- 0.05/2, df=4)
```

```
## [1] 2.776445
```

is bigger than...

```
qnorm(1- 0.05/2)
```

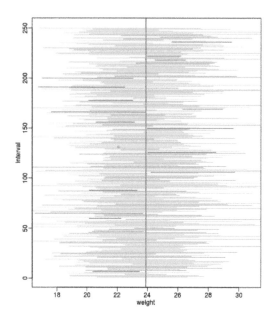

FIGURE 2.12
We show 250 random realizations of 95% confidence intervals, but now for a smaller sample size. The confidence is now based on the t-distribution approximation. The color denotes if the interval fell on the parameter or not.

```
## [1] 1.959964
```

...which makes the intervals larger and hence cover μ_X more frequently; in fact, about 95% of the time.

Connection Between Confidence Intervals and p-values We recommend that in practice confidence intervals be reported instead of p-values. If for some reason you are required to provide p-values, or required that your results are significant at the 0.05 of 0.01 levels, confidence intervals do provide this information.

If we are talking about a t-test p-value, we are asking if differences as extreme as the one we observe, $\bar{Y} - \bar{X}$, are likely when the difference between the population averages is actually equal to zero. So we can form a confidence interval with the observed difference. Instead of writing $\bar{Y} - \bar{X}$ repeatedly, let's define this difference as a new variable $d \equiv \bar{Y} - \bar{X}$.

Suppose you use CLT and report $d \pm 2s_d/\sqrt{N}$ as a 95% confidence interval for the difference and this interval does not include 0 (a false positive). Because the interval does not include 0, this implies that either $D - 2s_d/\sqrt{N} > 0$ or $d + 2s_d/\sqrt{N} < 0$. This suggests that either $\sqrt{N}d/s_d > 2$ or $\sqrt{N}d/s_d < 2$. This then implies that the t-statistic is more extreme than 2, which in turn suggests that the p-value must be smaller than 0.05 (approximately, for a more exact calculation use `qnorm(.05/2)` instead of 2). The same calculation can be made if we use the t-distribution instead of CLT (with `qt(.05/2, df=N-2)`). In summary, if a 95% or 99% confidence interval does not include 0, then the p-value must be smaller than 0.05 or 0.01 respectively.

Note that the confidence interval for the difference d is provided by the `t.test` function:

```
## [1] -0.04296563  6.08463229
## attr(,"conf.level")
```

```
## [1] 0.95
```

In this case, the 95% confidence interval does include 0 and we observe that the p-value is larger than 0.05 as predicted. If we change this to a 90% confidence interval, then:

```
t.test(treatment,control,conf.level=0.9)$conf.int
```

```
## [1] 0.4871597 5.5545070
## attr(,"conf.level")
## [1] 0.9
```

0 is no longer in the confidence interval (which is expected because the p-value is smaller than 0.10).

2.17   Power Calculations

Introduction We have used the example of the effects of two different diets on the weight of mice. Since in this illustrative example we have access to the population, we know that in fact there is a substantial (about 10%) difference between the average weights of the two populations:

```
library(dplyr)
dat <- read.csv("mice_pheno.csv") #Previously downloaded

controlPopulation <- filter(dat,Sex == "F" & Diet == "chow") %>%
  select(Bodyweight) %>% unlist

hfPopulation <- filter(dat,Sex == "F" & Diet == "hf") %>%
  select(Bodyweight) %>% unlist

mu_hf <- mean(hfPopulation)
mu_control <- mean(controlPopulation)
print(mu_hf - mu_control)
```

```
## [1] 2.375517
```

```
print((mu_hf - mu_control)/mu_control * 100) #percent increase
```

```
## [1] 9.942157
```

We have also seen that, in some cases, when we take a sample and perform a t-test, we don't always get a p-value smaller than 0.05. For example, here is a case where we take a sample of 5 mice and don't achieve statistical significance at the 0.05 level:

```
set.seed(1)
N <- 5
hf <- sample(hfPopulation,N)
control <- sample(controlPopulation,N)
t.test(hf,control)$p.value
```

```
## [1] 0.1410204
```

Did we make a mistake? By not rejecting the null hypothesis, are we saying the diet has no effect? The answer to this question is no. All we can say is that we did not reject the null hypothesis. But this does not necessarily imply that the null is true. The problem is that, in this particular instance, we don't have enough *power*, a term we are now going to define. If you are doing scientific research, it is very likely that you will have to do a power calculation at some point. In many cases, it is an ethical obligation as it can help you avoid sacrificing mice unnecessarily or limiting the number of human subjects exposed to potential risk in a study. Here we explain what statistical power means.

Types of Error Whenever we perform a statistical test, we are aware that we may make a mistake. This is why our p-values are not 0. Under the null, there is always a positive, perhaps very small, but still positive chance that we will reject the null when it is true. If the p-value is 0.05, it will happen 1 out of 20 times. This *error* is called *type I error* by statisticians.

A type I error is defined as rejecting the null when we should not. This is also referred to as a false positive. So why do we then use 0.05? Shouldn't we use 0.000001 to be really sure? The reason we don't use infinitesimal cut-offs to avoid type I errors at all cost is that there is another error we can commit: to not reject the null when we should. This is called a *type II error* or a false negative. The R code analysis above shows an example of a false negative: we did not reject the null hypothesis (at the 0.05 level) and, because we happen to know and peeked at the true population means, we know there is in fact a difference. Had we used a p-value cutoff of 0.25, we would not have made this mistake. However, in general, are we comfortable with a type I error rate of 1 in 4? Usually we are not.

The 0.05 and 0.01 Cut-offs Are Arbitrary Most journals and regulatory agencies frequently insist that results be significant at the 0.01 or 0.05 levels. Of course there is nothing special about these numbers other than the fact that some of the first papers on p-values used these values as examples. Part of the goal of this book is to give readers a good understanding of what p-values and confidence intervals are so that these choices can be judged in an informed way. Unfortunately, in science, these cut-offs are applied somewhat mindlessly, but that topic is part of a complicated debate.

Power Calculation Power is the probability of rejecting the null when the null is false. Of course "when the null is false" is a complicated statement because it can be false in many ways. $\Delta \equiv \mu_Y - \mu_X$ could be anything and the power actually depends on this parameter. It also depends on the standard error of your estimates which in turn depends on the sample size and the population standard deviations. In practice, we don't know these so we usually report power for several plausible values of Δ, σ_X, σ_Y and various sample sizes. Statistical theory gives us formulas to calculate power. The `pwr` package performs these calculations for you. Here we will illustrate the concepts behind power by coding up simulations in R.

Suppose our sample size is:

```
N <- 12
```

and we will reject the null hypothesis at:

```
alpha <- 0.05
```

What is our power with this particular data? We will compute this probability by re-running the exercise many times and calculating the proportion of times the null hypothesis is rejected. Specifically, we will run:

```
B <- 2000
```

simulations. The simulation is as follows: we take a sample of size N from both control and treatment groups, we perform a t-test comparing these two, and report if the p-value is less than `alpha` or not. We write a function that does this:

```
reject <- function(N, alpha=0.05){
   hf <- sample(hfPopulation,N)
   control <- sample(controlPopulation,N)
   pval <- t.test(hf,control)$p.value
   pval < alpha
}
```

Here is an example of one simulation for a sample size of 12. The call to `reject` answers the question "Did we reject?"

```
reject(12)
```

```
## [1] FALSE
```

Now we can use the `replicate` function to do this B times.

```
rejections <- replicate(B,reject(N))
```

Our power is just the proportion of times we correctly reject. So with $N = 12$ our power is only:

```
mean(rejections)
```

```
## [1] 0.2145
```

This explains why the t-test was not rejecting when we knew the null was false. With a sample size of just 12, our power is about 23%. To guard against false positives at the 0.05 level, we had set the threshold at a high enough level that resulted in many type II errors.

Let's see how power improves with N. We will use the function `sapply`, which applies a function to each of the elements of a vector. We want to repeat the above for the following sample size:

```
Ns <- seq(5, 50, 5)
```

So we use `apply` like this:

```
power <- sapply(Ns,function(N){
  rejections <- replicate(B, reject(N))
  mean(rejections)
  })
```

For each of the three simulations, the above code returns the proportion of times we reject. Not surprisingly power increases with N:

```
plot(Ns, power, type="b")
```

Similarly, if we change the level `alpha` at which we reject, power changes. The smaller I want the chance of type I error to be, the less power I will have. Another way of saying this is that we trade off between the two types of error. We can see this by writing similar code, but keeping N fixed and considering several values of `alpha`:

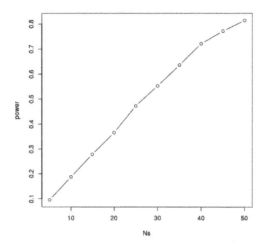

FIGURE 2.13
Power plotted against sample size.

```
N <- 30
alphas <- c(0.1,0.05,0.01,0.001,0.0001)
power <- sapply(alphas,function(alpha){
  rejections <- replicate(B,reject(N,alpha=alpha))
  mean(rejections)
})
plot(alphas, power, xlab="alpha", type="b", log="x")
```

Note that the x-axis in this last plot is in the log scale.

There is no "right" power or "right" alpha level, but it is important that you understand what each means.

To see this clearly, you could create a plot with curves of power versus N. Show several curves in the same plot with color representing alpha level.

p-values Are Arbitrary under the Alternative Hypothesis Another consequence of what we have learned about power is that p-values are somewhat arbitrary when the null hypothesis is not true and therefore the *alternative* hypothesis is true (the difference between the population means is not zero). When the alternative hypothesis is true, we can make a p-value as small as we want simply by increasing the sample size (supposing that we have an infinite population to sample from). We can show this property of p-values by drawing larger and larger samples from our population and calculating p-values. This works because, in our case, we know that the alternative hypothesis is true, since we have access to the populations and can calculate the difference in their means.

First write a function that returns a p-value for a given sample size N:

```
calculatePvalue <- function(N) {
  hf <- sample(hfPopulation,N)
  control <- sample(controlPopulation,N)
  t.test(hf,control)$p.value
}
```

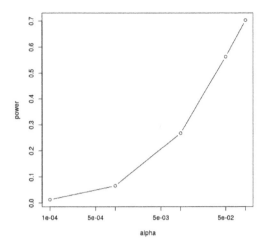

FIGURE 2.14
Power plotted against cut-off.

We have a limit here of 200 for the high-fat diet population, but we can see the effect well before we get to 200. For each sample size, we will calculate a few p-values. We can do this by repeating each value of N a few times.

```
Ns <- seq(10,200,by=10)
Ns_rep <- rep(Ns, each=10)
```

Again we use `sapply` to run our simulations:

```
pvalues <- sapply(Ns_rep, calculatePvalue)
```

Now we can plot the 10 p-values we generated for each sample size:

```
plot(Ns_rep, pvalues, log="y", xlab="sample size",
    ylab="p-values")
abline(h=c(.01, .05), col="red", lwd=2)
```

Note that the y-axis is log scale and that the p-values show a decreasing trend all the way to 10^{-8} as the sample size gets larger. The standard cutoffs of 0.01 and 0.05 are indicated with horizontal red lines.

It is important to remember that p-values are not more interesting as they become very very small. Once we have convinced ourselves to reject the null hypothesis at a threshold we find reasonable, having an even smaller p-value just means that we sampled more mice than was necessary. Having a larger sample size does help to increase the precision of our estimate of the difference Δ, but the fact that the p-value becomes very very small is just a natural consequence of the mathematics of the test. The p-values get smaller and smaller with increasing sample size because the numerator of the t-statistic has $\sqrt{N}$ (for equal sized groups, and a similar effect occurs when $M \neq N$). Therefore, if Δ is non-zero, the t-statistic will increase with N.

Therefore, a better statistic to report is the effect size with a confidence interval or some statistic which gives the reader a sense of the change in a meaningful scale. We can report the effect size as a percent by dividing the difference and the confidence interval by the control population mean:

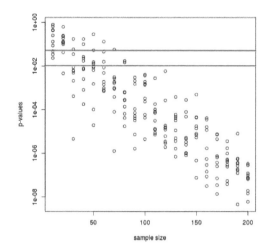

FIGURE 2.15

p-values from random samples at varying sample size. The actual value of the p-values decreases as we increase sample size whenever the alternative hypothesis is true.

```
N <- 12
hf <- sample(hfPopulation, N)
control <- sample(controlPopulation, N)
diff <- mean(hf) - mean(control)
diff / mean(control) * 100
```

```
## [1] 1.868663
```

```
t.test(hf, control)$conf.int / mean(control) * 100
```

```
## [1] -20.94576  24.68309
## attr(,"conf.level")
## [1] 0.95
```

In addition, we can report a statistic called Cohen's d[5], which is the difference between the groups divided by the pooled standard deviation of the two groups.

```
sd_pool <- sqrt(((N-1)*var(hf) + (N-1)*var(control))/(2*N - 2))
diff / sd_pool
```

```
## [1] 0.07140083
```

This tells us how many standard deviations of the data the mean of the high-fat diet group is from the control group. Under the alternative hypothesis, unlike the t-statistic which is guaranteed to increase, the effect size and Cohen's d will become more precise.

---

[5]https://en.wikipedia.org/wiki/Effect_size#Cohen.27s_d

2.18   Exercises

For these exercises we will load the babies dataset from `babies.txt`. We will use this data to review the concepts behind the p-values and then test confidence interval concepts.

```
url <- "https://raw.githubusercontent.com/genomicsclass/dagdata/master/inst/extdata/babies.txt"
filename <- basename(url)
download(url, destfile=filename)
babies <- read.table("babies.txt", header=TRUE)
```

This is a large dataset (1,236 cases), and we will pretend that it contains the entire population in which we are interested. We will study the differences in birth weight between babies born to smoking and non-smoking mothers.

First, let's split this into two birth weight datasets: one of birth weights to non-smoking mothers and the other of birth weights to smoking mothers.

```
bwt.nonsmoke <- filter(babies, smoke==0) %>% select(bwt) %>% unlist
bwt.smoke <- filter(babies, smoke==1) %>% select(bwt) %>% unlist
```

Now, we can look for the true population difference in means between smoking and non-smoking birth weights.

```
library(rafalib)
mean(bwt.nonsmoke)-mean(bwt.smoke)
popsd(bwt.nonsmoke)
popsd(bwt.smoke)
```

The population difference of mean birth weights is about 8.9 ounces. The standard deviations of the nonsmoking and smoking groups are about 17.4 and 18.1 ounces, respectively.

As we did with the mouse weight data, this assessment interactively reviews inference concepts using simulations in R. We will treat the babies dataset as the full population and draw samples from it to simulate individual experiments. We will then ask whether somebody who only received the random samples would be able to draw correct conclusions about the population.

We are interested in testing whether the birth weights of babies born to non-smoking mothers are significantly different from the birth weights of babies born to smoking mothers.

1. Set the seed at 1 and obtain two samples, each of size $N = 25$, from non-smoking mothers (`dat.ns`) and smoking mothers (`dat.s`). Compute the t-statistic (call it `tval`).

2. Recall that we summarize our data using a t-statistics because we know that in situations where the null hypothesis is true (what we mean when we say "under the null") and the sample size is relatively large, this t-value will have an approximate standard normal distribution. Because we know the distribution of the t-value under the null, we can quantitatively determine how unusual the observed t-value would be if the null hypothesis were true.

The standard procedure is to examine the probability a t-statistic that actually does follow the null hypothesis would have larger absolute value than the absolute value of the t-value we just observed – this is called a two-sided test.

We have computed these by taking one minus the area under the standard normal curve between `-abs(tval)` and `abs(tval)`. In R, we can do this by using the `pnorm` function, which computes the area under a normal curve from negative infinity up to the value given as its first argument:

3. Because of the symmetry of the standard normal distribution, there is a simpler way to calculate the probability that a t-value under the null could have a larger absolute value than `tval`. Choose the simplified calculation from the following:

   (a) `1-2*pnorm(abs(tval))`

   (b) `1-2*pnorm(-abs(tval))`

   (c) `1-pnorm(-abs(tval))`

   (d) `2*pnorm(-abs(tval))`

4. By reporting only p-values, many scientific publications provide an incomplete story of their findings. As we have mentioned, with very large sample sizes, scientifically insignificant differences between two groups can lead to small p-values. Confidence intervals are more informative as they include the estimate itself. Our estimate of the difference between babies of smokers and non-smokers: `mean(dat.s) - mean( dat.ns)`. If we use the CLT, what quantity would we add and subtract to this estimate to obtain a 99% confidence interval?

5. If instead of CLT, we use the t-distribution approximation, what do we add and subtract (use `2*N-2` degrees of freedom)?

6. Why are the values from 4 and 5 so similar?

   (a) Coincidence.

   (b) They are both related to 99% confidence intervals.

   (c) `N` and thus the degrees of freedom is large enough to make the normal and t-distributions very similar.

   (d) They are actually quite different, differing by more than 1 ounce.

7. No matter which way you compute it, the p-value `pval` is the probability that the null hypothesis could have generated a t-statistic more extreme than what we observed: `tval`. If the p-value is very small, this means that observing a value more extreme than `tval` would be very rare if the null hypothesis were true, and would give strong evidence that we should **reject** the null hypothesis. We determine how small the p-value needs to be to reject the null by deciding how often we would be willing to mistakenly reject the null hypothesis.

The standard decision rule is the following: choose some small value α (in most disciplines the conventional choice is $\alpha = 0.05$) and reject the null hypothesis if the p-value is less than α. We call α the *significance level* of the test.

It turns out that if we follow this decision rule, the probability that we will reject the null hypothesis by mistake is equal to α. (This fact is not immediately obvious and requires some probability theory to show.) We call the *event* of rejecting the null hypothesis, when it is in fact true, a *Type I error*, we call the *probability* of making a Type I error, the *Type I error rate*, and we say that rejecting the null hypothesis when the p-value is less than α, *controls* the Type I error rate so that it is equal to α. We will see a number of decision rules that we use in order to control the probabilities of other types of errors. Often, we will guarantee that the probability of an error is less than some level, but, in this case, we can guarantee that the probability of a Type I error is *exactly equal* to α.

Which of the following sentences about a Type I error is **not** true?

- A) The following is another way to describe a Type I error: you decided to reject the null hypothesis on the basis of data that was actually generated by the null hypothesis.

- B) The following is another way to describe a Type I error: due to random fluctuations, even though the data you observed were actually generated by the null hypothesis, the p-value calculated from the observed data was small, so you rejected it.

- C) From the original data alone, you can tell whether you have made a Type I error.

- D) In scientific practice, a Type I error constitutes reporting a "significant" result when there is actually no result.

8. In the simulation we have set up here, we know the null hypothesis is false – the true value of difference in means is actually around 8.9. Thus, we are concerned with how often the decision rule outlined in the last section allows us to conclude that the null hypothesis is actually false. In other words, we would like to quantify the *Type II error rate* of the test, or the probability that we fail to reject the null hypothesis when the alternative hypothesis is true.

Unlike the Type I error rate, which we can characterize by assuming that the null hypothesis of "no difference" is true, the Type II error rate cannot be computed by assuming the alternative hypothesis alone because the alternative hypothesis alone does not specify a particular value for the difference. It thus does not nail down a specific distribution for the t-value under the alternative.

For this reason, when we study the Type II error rate of a hypothesis testing procedure, we need to assume a particular *effect size*, or hypothetical size of the difference between population means, that we wish to target. We ask questions such as "what is the smallest difference I could reliably distinguish from 0 given my sample size N?" or, more commonly, "How big does N have to be in order to detect that the absolute value of the difference is greater than zero?" Type II error control plays a major role in designing data collection procedures **before** you actually see the data, so that you know the test you will run has enough sensitivity or *power*. Power is one minus the Type II error rate, or the probability that you will reject the null hypothesis when the alternative hypothesis is true.

There are several aspects of a hypothesis test that affect its power for a particular effect size. Intuitively, setting a lower α decreases the power of the test for a given effect size because the null hypothesis will be more difficult to reject. This means that for an experiment with fixed parameters (i.e., with a predetermined sample size, recording mechanism, etc), the power of the hypothesis test trades off with its Type I error rate, no matter what effect size you target.

We can explore the trade off of power and Type I error concretely using the babies data. Since we have the full population, we know what the true effect size is (about 8.93) and we can compute the power of the test for true difference between populations.

Set the seed at 1 and take a random sample of $N = 5$ measurements from each of the smoking and nonsmoking datasets. What is the p-value (use the `t-test` function)?

9. The p-value is larger than 0.05 so using the typical cut-off, we would not reject. This is a type II error. Which of the following is *not* a way to decrease this type of error?

   (a) Increase our chance of a type I error.
   (b) Take a larger sample size.
   (c) Find a population for which the null is not true.
   (d) Use a higher α level.

10. Set the seed at 1, then use the `replicate` function to repeat the code used in exercise 9 10,000 times. What proportion of the time do we reject at the 0.05 level?

11. Note that, not surprisingly, the power is lower than 10%. Repeat the exercise above for sample sizes of 30, 60, 90 and 120. Which of those four gives you power of about 80%?

12. Repeat problem 11, but now require an α level of 0.01. Which of those four gives you power of about 80%?

2.19   Monte Carlo Simulation

Computers can be used to generate pseudo-random numbers. For practical purposes these pseudo-random numbers can be used to imitate random variables from the real world. This permits us to examine properties of random variables using a computer instead of theoretical or analytical derivations. One very useful aspect of this concept is that we can create *simulated* data to test out ideas or competing methods, without actually having to perform laboratory experiments.

Simulations can also be used to check theoretical or analytical results. Also, many of the theoretical results we use in statistics are based on asymptotics: they hold when the sample size goes to infinity. In practice, we never have an infinite number of samples so we may want to know how well the theory works with our actual sample size. Sometimes we can answer this question analytically, but not always. Simulations are extremely useful in these cases.

As an example, let's use a Monte Carlo simulation to compare the CLT to the t-distribution approximation for different sample sizes.

```
library(dplyr)
dat <- read.csv("mice_pheno.csv")
controlPopulation <- filter(dat,Sex == "F" & Diet == "chow") %>%
  select(Bodyweight) %>% unlist
```

We will build a function that automatically generates a t-statistic under the null hypothesis for a sample size of **n**.

```
ttestgenerator <- function(n) {
  #note that here we have a false "high fat" group where we actually
  #sample from the chow or control population.
  #This is because we are modeling the null.
  cases <- sample(controlPopulation,n)
  controls <- sample(controlPopulation,n)
  tstat <- (mean(cases)-mean(controls)) /
      sqrt( var(cases)/n + var(controls)/n )
  return(tstat)
  }
ttests <- replicate(1000, ttestgenerator(10))
```

With 1,000 Monte Carlo simulated occurrences of this random variable, we can now get a glimpse of its distribution:

```
hist(ttests)
```

So is the distribution of this t-statistic well approximated by the normal distribution? In the next chapter, we will formally introduce quantile-quantile plots, which provide a useful

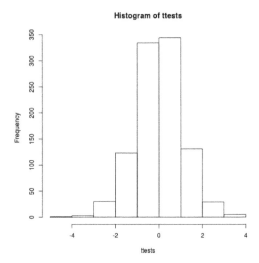

FIGURE 2.16
Histogram of 1000 Monte Carlo simulated t-statistics.

visual inspection of how well one distribution approximates another. As we will explain later, if points fall on the identity line, it means the approximation is a good one.

```
qqnorm(ttests)
abline(0,1)
```

This looks like a very good approximation. For this particular population, a sample size of 10 was large enough to use the CLT approximation. How about 3?

```
ttests <- replicate(1000, ttestgenerator(3))
qqnorm(ttests)
abline(0,1)
```

Now we see that the large quantiles, referred to by statisticians as the *tails*, are larger than expected (below the line on the left side of the plot and above the line on the right side of the plot). In the previous module, we explained that when the sample size is not large enough and the *population values* follow a normal distribution, then the t-distribution is a better approximation. Our simulation results seem to confirm this:

```
ps <- (seq(0,999)+0.5)/1000
qqplot(qt(ps,df=2*3-2),ttests,xlim=c(-6,6),ylim=c(-6,6))
abline(0,1)
```

The t-distribution is a much better approximation in this case, but it is still not perfect. This is due to the fact that the original data is not that well approximated by the normal distribution.

```
qqnorm(controlPopulation)
qqline(controlPopulation)
```

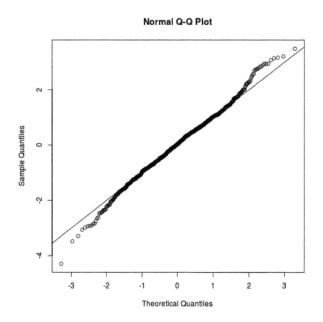

FIGURE 2.17
Quantile-quantile plot comparing 1000 Monte Carlo simulated t-statistics to theoretical
normal distribution.

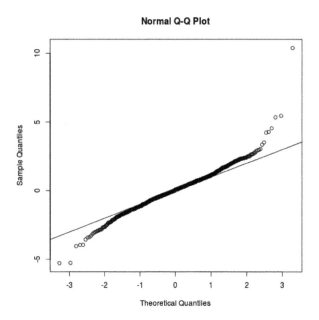

FIGURE 2.18
Quantile-quantile plot comparing 1000 Monte Carlo simulated t-statistics with three degrees
of freedom to theoretical normal distribution.

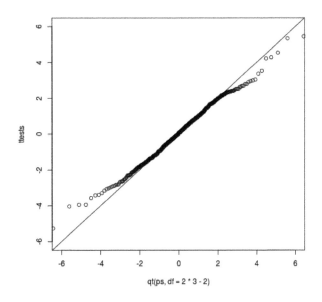

FIGURE 2.19
Quantile-quantile plot comparing 1000 Monte Carlo simulated t-statistics with three degrees of freedom to theoretical t-distribution.

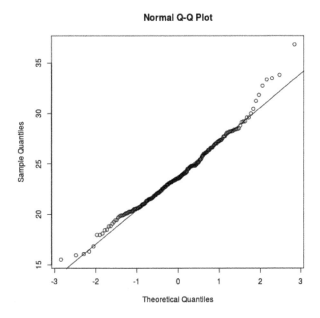

FIGURE 2.20
Quantile-quantile of original data compared to theoretical quantile distribution.

2.20   Parametric Simulations for the Observations

The technique we used to motivate random variables and the null distribution was a type of Monte Carlo simulation. We had access to population data and generated samples at random. In practice, we do not have access to the entire population. The reason for using the approach here was for educational purposes. However, when we want to use Monte Carlo simulations in practice, it is much more typical to assume a parametric distribution and generate a population from this, which is called a *parametric simulation*. This means that we take parameters estimated from the real data (here the mean and the standard deviation), and plug these into a model (here the normal distribution). This is actually the most common form of Monte Carlo simulation.

For the case of weights, we could use our knowledge that mice typically weigh 24 grams with a SD of about 3.5 grams, and that the distribution is approximately normal, to generate population data:

```
controls<- rnorm(5000, mean=24, sd=3.5)
```

After we generate the data, we can then repeat the exercise above. We no longer have to use the `sample` function since we can re-generate random normal numbers. The `ttestgenerator` function therefore can be written as follows:

```
ttestgenerator <- function(n, mean=24, sd=3.5) {
  cases <- rnorm(n,mean,sd)
  controls <- rnorm(n,mean,sd)
  tstat <- (mean(cases)-mean(controls)) /
      sqrt( var(cases)/n + var(controls)/n )
  return(tstat)
}
```

2.21   Exercises

We have used Monte Carlo simulation throughout this chapter to demonstrate statistical concepts; namely, sampling from the population. We mostly applied this to demonstrate the statistical properties related to inference on differences in averages. Here, we will consider examples of how Monte Carlo simulations are used in practice.

1. Imagine you are William Sealy Gosset[6] and have just mathematically derived the distribution of the t-statistic when the sample comes from a normal distribution. Unlike Gosset you have access to computers and can use them to check the results.

    Let's start by creating an outcome. Set the seed at 1, use `rnorm` to generate a random sample of size 5, $X_1, \ldots, X_5$ from a standard normal distribution, then compute the t-statistic $t = \sqrt{5}\,\bar{X}/s$ with s the sample standard deviation. What value do you observe?

2. You have just performed a Monte Carlo simulation using `rnorm`, a random number generator for normally distributed data. Gosset's mathematical calculation tells us that the t-statistic defined in the previous exercise, a random variable, follows

---

[6]https://en.wikipedia.org/wiki/William_Sealy_Gosset

a t-distribution with $N - 1$ degrees of freedom. Monte Carlo simulations can be used to check the theory: we generate many outcomes and compare them to the theoretical result. Set the seed to 1, generate $B = 1000$ t-statistics as done in exercise 1. What proportion is larger than 2?

3. The answer to exercise 2 is very similar to the theoretical prediction: `1-pt(2,df=4)`. We can check several such quantiles using the `qqplot` function.

To obtain quantiles for the t-distribution we can generate percentiles from just above 0 to just below 1: `B=100; ps = seq(1/(B+1), 1-1/(B+1),len=B)` and compute the quantiles with `qt(ps,df=4)`. Now we can use `qqplot` to compare these theoretical quantiles to those obtained in the Monte Carlo simulation. Use Monte Carlo simulation developed for exercise 2 to corroborate that the t-statistic $t = \sqrt{N} \bar{X}/s$ follows a t-distribution for several values of N.

For which sample sizes does the approximation best work?

- Larger sample sizes.

- Smaller sample sizes.

- The approximations are spot on for all sample sizes.

- None. We should use CLT instead.

4. Use Monte Carlo simulation to corroborate that the t-statistic comparing two means and obtained with normally distributed (mean 0 and sd) data follows a t-distribution. In this case we will use the `t.test` function with `var.equal=TRUE`. With this argument the degrees of freedom will be `df=2*N-2` with N the sample size. For which sample sizes does the approximation best work?
   (a) Larger sample sizes.
   (b) Smaller sample sizes.
   (c) The approximations are spot on for all sample sizes.
   (d) None. We should use CLT instead.

5. Is the following statement true or false? If instead of generating the sample with `X=rnorm(15)`, we generate it with binary data (either positive or negative 1 with probability 0.5) `X =sample(c(-1,1), 15, replace=TRUE)` then the t-statistic

```
tstat <- sqrt(15)*mean(X) / sd(X)
```

is approximated by a t-distribution with 14 degrees of freedom.

6. Is the following statement true or false? If instead of generating the sample with `X=rnorm(N)` with N=1000, we generate the data with binary data `X= sample(c(-1,1), N, replace=TRUE)`, then the t-statistic `sqrt(N)*mean(X)/sd(X)` is approximated by a t-distribution with 999 degrees of freedom.

7. We can derive approximation of the distribution of the sample average or the t-statistic theoretically. However, suppose we are interested in the distribution of a statistic for which a theoretical approximation is not immediately obvious.

Consider the sample median as an example. Use a Monte Carlo to determine which of the following best approximates the median of a sample taken from normally distributed population with mean 0 and standard deviation 1.

- Just like for the average, the sample median is approximately normal with mean 0 and SD $1/\sqrt{N}$.

- The sample median is not approximately normal.

- The sample median is t-distributed for small samples and normally distributed for large ones.

- The sample median is approximately normal with mean 0 and SD larger than $1/\sqrt{N}$.

2.22   Permutation Tests

Suppose we have a situation in which none of the standard mathematical statistical approximations apply. We have computed a summary statistic, such as the difference in mean, but do not have a useful approximation, such as that provided by the CLT. In practice, we do not have access to all values in the population so we can't perform a simulation as done above. Permutation tests can be useful in these scenarios.

We are back to the scenario where we only have 10 measurements for each group.

```
dat=read.csv("femaleMiceWeights.csv")
```

```
library(dplyr)
```

```
control <- filter(dat,Diet=="chow") %>% select(Bodyweight) %>% unlist
treatment <- filter(dat,Diet=="hf") %>% select(Bodyweight) %>% unlist
obsdiff <- mean(treatment)-mean(control)
```

In previous sections, we showed parametric approaches that helped determine if the observed difference was significant. Permutation tests take advantage of the fact that if we randomly shuffle the cases and control labels, then the null is true. So we shuffle the cases and control labels and assume that the ensuing distribution approximates the null distribution. Here is how we generate a null distribution by shuffling the data 1,000 times:

```
N <- 12
avgdiff <- replicate(1000, {
    all <- sample(c(control,treatment))
    newcontrols <- all[1:N]
    newtreatments <- all[(N+1):(2*N)]
  return(mean(newtreatments) - mean(newcontrols))
})
hist(avgdiff)
abline(v=obsdiff, col="red", lwd=2)
```

How many of the null means are bigger than the observed value? That proportion would be the p-value for the null. We add a 1 to the numerator and denominator to account for misestimation of the p-value (for more details see Phipson and Smyth, Permutation P-values should never be zero[7]).

```
#the proportion of permutations with larger difference
(sum(abs(avgdiff) > abs(obsdiff)) + 1) / (length(avgdiff) + 1)
```

```
## [1] 0.07392607
```

---

[7]http://www.ncbi.nlm.nih.gov/pubmed/21044043

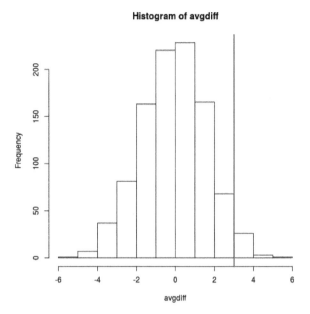

FIGURE 2.21
Histogram of difference between averages from permutations. Vertical line shows the observed difference.

Now let's repeat this experiment for a smaller dataset. We create a smaller dataset by sampling:

```
N <- 5
control <- sample(control,N)
treatment <- sample(treatment,N)
obsdiff <- mean(treatment)- mean(control)
```

and repeat the exercise:

```
avgdiff <- replicate(1000, {
    all <- sample(c(control,treatment))
    newcontrols <- all[1:N]
    newtreatments <- all[(N+1):(2*N)]
  return(mean(newtreatments) - mean(newcontrols))
})
hist(avgdiff)
abline(v=obsdiff, col="red", lwd=2)
```

Now the observed difference is not significant using this approach. Keep in mind that there is no theoretical guarantee that the null distribution estimated from permutations approximates the actual null distribution. For example, if there is a real difference between the populations, some of the permutations will be unbalanced and will contain some samples that explain this difference. This implies that the null distribution created with permutations will have larger tails than the actual null distribution. This is why permutations result in conservative p-values. For this reason, when we have few samples, we can't do permutations.

Note also that permutation tests still have assumptions: samples are assumed to be

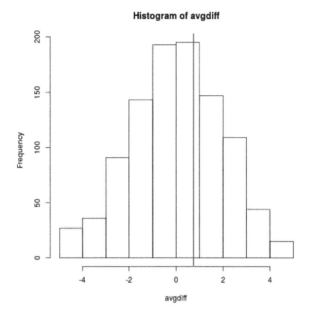

FIGURE 2.22

Histogram of difference between averages from permutations for smaller sample size. Vertical line shows the observed difference.

independent and "exchangeable". If there is hidden structure in your data, then permutation tests can result in estimated null distributions that underestimate the size of tails because the permutations may destroy the existing structure in the original data.

2.23   Exercises

We will use the following dataset to demonstrate the use of permutations:

```
dir <- "https://raw.githubusercontent.com/genomicsclass/dagdata/master/inst/extdata/"
filename <- "babies.txt"
url <- paste0(dir, filename)
download(url, destfile=filename)
babies <- read.table("babies.txt", header=TRUE)
bwt.nonsmoke <- filter(babies, smoke==0) %>% select(bwt) %>% unlist
bwt.smoke <- filter(babies, smoke==1) %>% select(bwt) %>% unlist
```

1. We will generate the following random variable based on a sample size of 10 and observe the following difference:

```
N=10
set.seed(1)
nonsmokers <- sample(bwt.nonsmoke , N)
smokers <- sample(bwt.smoke , N)
obs <- mean(smokers) - mean(nonsmokers)
```

The question is whether this observed difference is statistically significant. We do not want to rely on the assumptions needed for the normal or t-distribution approximations to hold, so instead we will use permutations. We will reshuffle the data and recompute the mean. We can create one permuted sample with the following code:

```
dat <- c(smokers,nonsmokers)
shuffle <- sample( dat )
smokersstar <- shuffle[1:N]
nonsmokersstar <- shuffle[(N+1):(2*N)]
mean(smokersstar)-mean(nonsmokersstar)
```

The last value is one observation from the null distribution we will construct. Set the seed at 1, and then repeat the permutation 1,000 times to create a null distribution. What is the permutation derived p-value for our observation?

2. Repeat the above exercise, but instead of the differences in mean, consider the differences in median obs <- median(smokers) - median(nonsmokers). What is the permutation based p-value?

2.24 Association Tests

The statistical tests we have covered up to now leave out a substantial portion of life science projects. Specifically, we are referring to data that is binary, categorical and ordinal. To give a very specific example, consider genetic data where you have two groups of genotypes (AA/Aa or aa) for cases and controls for a given disease. The statistical question is if genotype and disease are associated. As in the examples we have been studying previously, we have two populations (AA/Aa and aa) and then numeric data for each, where disease status can be coded as 0 or 1. So why can't we perform a t-test? Note that the data is either 0 (control) or 1 (cases). It is pretty clear that this data is not normally distributed so the t-distribution approximation is certainly out of the question. We could use CLT if the sample size is large enough; otherwise, we can use *association tests*.

Lady Tasting Tea One of the most famous examples of hypothesis testing was performed by R.A. Fisher[8]. An acquaintance of Fisher's claimed that she could tell if milk was added before or after tea was poured. Fisher gave her four pairs of cups of tea: one with milk poured first, the other after. The order was randomized. Say she picked 3 out of 4 correctly, do we believe she has a special ability? Hypothesis testing helps answer this question by quantifying what happens by chance. This example is called the "Lady Tasting Tea" experiment (and, as it turns out, Fisher's friend was a scientist herself, Muriel Bristol[9]).

The basic question we ask is: if the tester is actually guessing, what are the chances that she gets 3 or more correct? Just as we have done before, we can compute a probability under the null hypothesis that she is guessing 4 of each. If we assume this null hypothesis, we can think of this particular example as picking 4 balls out of an urn with 4 green (correct answer) and 4 red (incorrect answer) balls.

Under the null hypothesis that she is simply guessing, each ball has the same chance of being picked. We can then use combinatorics to figure out each probability. The probability of picking 3 is $\binom{4}{3}\binom{4}{1}/\binom{8}{4} = 16/70$. The probability of picking all 4 correct is

[8]https://en.wikipedia.org/wiki/Ronald_Fisher
[9]https://en.wikipedia.org/wiki/Muriel_Bristol

$\binom{4}{4}\binom{4}{0}/\binom{8}{4} = 1/70$. Thus, the chance of observing a 3 or something more extreme, under the null hypothesis, is ≈ 0.24. This is the p-value. The procedure that produced this p-value is called *Fisher's exact test* and it uses the *hypergeometric distribution*.

Two By Two Tables The data from the experiment above can be summarized by a two by two table:

```
tab <- matrix(c(3,1,1,3),2,2)
rownames(tab)<-c("Poured Before","Poured After")
colnames(tab)<-c("Guessed before","Guessed after")
tab
```

```
##                Guessed before Guessed after
## Poured Before               3             1
## Poured After                1             3
```

The function `fisher.test` performs the calculations above and can be obtained like this:

```
fisher.test(tab,alternative="greater")
```

```
##
##  Fisher's Exact Test for Count Data
##
## data:  tab
## p-value = 0.2429
## alternative hypothesis: true odds ratio is greater than 1
## 95 percent confidence interval:
##   0.3135693        Inf
## sample estimates:
## odds ratio
##    6.408309
```

Chi-square Test Genome-wide association studies (GWAS) have become ubiquitous in biology. One of the main statistical summaries used in these studies are Manhattan plots. The y-axis of a Manhattan plot typically represents the negative of log (base 10) of the p-values obtained for association tests applied at millions of single nucleotide polymorphisms (SNP). The x-axis is typically organized by chromosome (chromosome 1 to 22, X, Y, etc.). These p-values are obtained in a similar way to the test performed on the tea taster. However, in that example the number of green and red balls is experimentally fixed and the number of answers given for each category is also fixed. Another way to say this is that the sum of the rows and the sum of the columns are fixed. This defines constraints on the possible ways we can fill the two by two table and also permits us to use the hypergeometric distribution. In general, this is not the case. Nonetheless, there is another approach, the Chi-squared test, which is described below.

Imagine we have 250 individuals, where some of them have a given disease and the rest do not. We observe that 20% of the individuals that are homozygous for the minor allele (aa) have the disease compared to 10% of the rest. Would we see this again if we picked another 250 individuals?

Let's create a dataset with these percentages:

```
disease=factor(c(rep(0,180),rep(1,20),rep(0,40),rep(1,10)),
```

```
            labels=c("control","cases"))
genotype=factor(c(rep("AA/Aa",200),rep("aa",50)),
              levels=c("AA/Aa","aa"))
dat <- data.frame(disease, genotype)
dat <- dat[sample(nrow(dat)),] #shuffle them up
head(dat)
```

```
##        disease genotype
## 67   control    AA/Aa
## 93   control    AA/Aa
## 143  control    AA/Aa
## 225  control       aa
## 50   control    AA/Aa
## 221  control       aa
```

To create the appropriate two by two table, we will use the function `table`. This function tabulates the frequency of each level in a factor. For example:

```
table(genotype)
```

```
## genotype
## AA/Aa    aa
##   200    50
```

```
table(disease)
```

```
## disease
## control   cases
##     220      30
```

If you provide the function with two factors, it will tabulate all possible pairs and thus create the two by two table:

```
tab <- table(genotype,disease)
tab
```

```
##           disease
## genotype control cases
##    AA/Aa     180    20
##       aa      40    10
```

Note that you can feed `table` n factors and it will tabulate all n-tables.

The typical statistics we use to summarize these results is the odds ratio (OR). We compute the odds of having the disease if you are an "aa": 10/40, the odds of having the disease if you are an "AA/Aa": 20/180, and take the ratio: (10/40)/(20/180)

```
(tab[2,2]/tab[2,1]) / (tab[1,2]/tab[1,1])
```

```
## [1] 2.25
```

To compute a p-value, we don't use the OR directly. We instead assume that there is no association between genotype and disease, and then compute what we expect to see in each *cell* of the table (note: this use of the word "cell" refers to elements in a matrix or table and has nothing to do with biological cells). Under the null hypothesis, the group with 200 individuals and the group with 50 individuals were each randomly assigned the disease with the same probability. If this is the case, then the probability of disease is:

```
p=mean(disease=="cases")
p
```

[1] 0.12

The expected table is therefore:

```
expected <- rbind(c(1-p,p)*sum(genotype=="AA/Aa"),
                  c(1-p,p)*sum(genotype=="aa"))
dimnames(expected)<-dimnames(tab)
expected
```

```
##          disease
## genotype control cases
##    AA/Aa     176    24
##       aa      44     6
```

The Chi-square test uses an asymptotic result (similar to the CLT) related to the sums of independent binary outcomes. Using this approximation, we can compute the probability of seeing a deviation from the expected table as big as the one we saw. The p-value for this table is:

```
chisq.test(tab)$p.value
```

[1] 0.08857435

Large Samples, Small p-values As mentioned earlier, reporting only p-values is not an appropriate way to report the results of your experiment. Many genetic association studies seem to overemphasize p-values. They have large sample sizes and report impressively small p-values. Yet when one looks closely at the results, we realize odds ratios are quite modest: barely bigger than 1. In this case the difference of having genotype AA/Aa or aa might not change an individual's risk for a disease in an amount which is *practically significant*, in that one might not change one's behavior based on the small increase in risk.

There is not a one-to-one relationship between the odds ratio and the p-value. To demonstrate, we recalculate the p-value keeping all the proportions identical, but increasing the sample size by 10, which reduces the p-value substantially (as we saw with the t-test under the alternative hypothesis):

```
tab<-tab*10
chisq.test(tab)$p.value
```

[1] 1.219624e-09

Confidence Intervals for the Odds Ratio Computing confidence intervals for the OR is not mathematically straightforward. Unlike other statistics, for which we can derive useful approximations of their distributions, the OR is not only a ratio, but a ratio of ratios. Therefore, there is no simple way of using, for example, the CLT.

One approach is to use the theory of *generalized linear models* which provides estimates of the log odds ratio, rather than the OR itself, that can be shown to be asymptotically normal. Here we provide R code without presenting the theoretical details (for further details please see a reference on generalized linear models such as Wikipedia[10] or McCullagh and Nelder, 1989[11]):

---

[10] https://en.wikipedia.org/wiki/Generalized_linear_model
[11] https://books.google.com/books?hl=en&lr=&id=h9kFH2_FfBkC

```
fit <- glm(disease~genotype,family="binomial",data=dat)
coeftab<- summary(fit)$coef
coeftab
```

```
##               Estimate Std. Error    z value      Pr(>|z|)
## (Intercept) -2.1972246  0.2356828 -9.322803 1.133070e-20
## genotypeaa   0.8109302  0.4249074  1.908487 5.632834e-02
```

The second row of the table shown above gives you the estimate and SE of the log odds ratio. Mathematical theory tells us that this estimate is approximately normally distributed. We can therefore form a confidence interval and then exponentiate to provide a confidence interval for the OR.

```
ci <- coeftab[2,1] + c(-2,2)*coeftab[2,2]
exp(ci)
```

```
## [1] 0.9618616 5.2632310
```

The confidence includes 1, which is consistent with the p-value being bigger than 0.05. Note that the p-value shown here is based on a different approximation to the one used by the Chi-square test, which is why they differ.

2.25 Exercises

We showed how to calculate a Chi-square test from a table. Here we will show how to generate the table from data which is in the form of a dataframe, so that you can then perform an association test to see if two columns have an enrichment (or depletion) of shared occurrences.

Download the https://studio.edx.org/c4x/HarvardX/PH525.1x/asset/assoctest.csv[12] file into your R working directory, and then read it into R: `d = read.csv("assoctest.csv")`

1. This dataframe reflects the allele status (either AA/Aa or aa) and the case/control status for 72 individuals. Compute the Chi-square test for the association of genotype with case/control status (using the `table` function and the `chisq.test` function). Examine the table to see if it looks enriched for association by eye. What is the X-squared statistic?

2. Compute Fisher's exact test `fisher.test` for the same table. What is the p-value?

---

[12]`assoctest.csv`

3

Exploratory Data Analysis

"The greatest value of a picture is when it forces us to notice what we never expected to see." -John W. Tukey

Biases, systematic errors and unexpected variability are common in data from the life sciences. Failure to discover these problems often leads to flawed analyses and false discoveries. As an example, consider that experiments sometimes fail and not all data processing pipelines, such as the t.test function in R, are designed to detect these. Yet, these pipelines still give you an answer. Furthermore, it may be hard or impossible to notice an error was made just from the reported results.

Graphing data is a powerful approach to detecting these problems. We refer to this as *exploratory data analysis* (EDA). Many important methodological contributions to existing techniques in data analysis were initiated by discoveries made via EDA. In addition, EDA can lead to interesting biological discoveries which would otherwise be missed through simply subjecting the data to a battery of hypothesis tests. Through this book, we make use of exploratory plots to motivate the analyses we choose. Here we present a general introduction to EDA using height data.

We have already introduced some EDA approaches for *univariate* data, namely the histograms and qq-plot. Here we describe the qq-plot in more detail and some EDA and summary statistics for paired data. We also give a demonstration of commonly used figures that we recommend against.

3.1 Quantile Quantile Plots

To corroborate that a theoretical distribution, for example the normal distribution, is in fact a good approximation, we can use quantile-quantile plots (qq-plots). Quantiles are best understood by considering the special case of percentiles. The p-th percentile of a list of a distribution is defined as the number q that is bigger than p% of numbers (so the inverse of the cumulative distribution function we defined earlier). For example, the median 50-th percentile is the median. We can compute the percentiles for our list of heights:

```
library(rafalib)
data(father.son,package="UsingR") ##available from CRAN
x <- father.son$fheight
```

and for the normal distribution:

```
ps <- ( seq(0,99) + 0.5 )/100
qs <- quantile(x, ps)
normalqs <- qnorm(ps, mean(x), popsd(x))
plot(normalqs,qs,xlab="Normal percentiles",ylab="Height percentiles")
abline(0,1) ##identity line
```

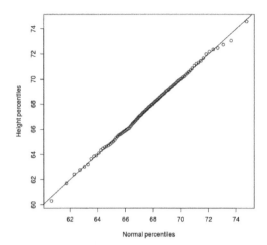

FIGURE 3.1

First example of qqplot. Here we compute the theoretical quantiles ourselves.

Note how close these values are. Also, note that we can see these qq-plots with less code (this plot has more points than the one we constructed manually, and so tail-behavior can be seen more clearly).

```
qqnorm(x)
qqline(x)
```

However, the `qqnorm` function plots against a standard normal distribution. This is why the line has slope `popsd(x)` and intercept `mean(x)`.

In the example above, the points match the line very well. In fact, we can run Monte Carlo simulations to see plots like this for data known to be normally distributed.

```
n <-1000
x <- rnorm(n)
qqnorm(x)
qqline(x)
```

We can also get a sense for how non-normally distributed data will look in a qq-plot. Here we generate data from the t-distribution with different degrees of freedom. Notice that the smaller the degrees of freedom, the fatter the tails. We call these "fat tails" because if we plotted an empirical density or histogram, the density at the extremes would be higher than the theoretical curve. In the qq-plot, this can be seen in that the curve is lower than the identity line on the left side and higher on the right side. This means that there are more extreme values than predicted by the theoretical density plotted on the x-axis.

```
dfs <- c(3,6,12,30)
mypar(2,2)
for(df in dfs){
  x <- rt(1000,df)
  qqnorm(x,xlab="t quantiles",main=paste0("d.f=",df),ylim=c(-6,6))
  qqline(x)
}
```

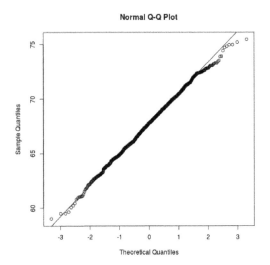

FIGURE 3.2
Second example of qqplot. Here we use the function qqnorm which computes the theoretical normal quantiles automatically.

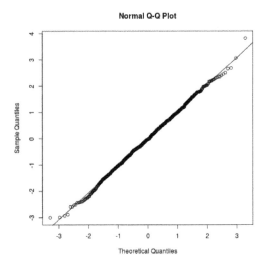

FIGURE 3.3
Example of the qqnorm function. Here we apply it to numbers generated to follow a normal distribution.

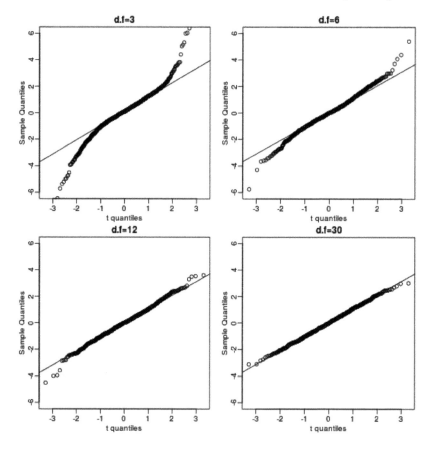

FIGURE 3.4
We generate t-distributed data for four degrees of freedom and make qqplots against normal
theoretical quantiles.

3.2   Boxplots

Data is not always normally distributed. Income is a widely cited example. In these cases,
the average and standard deviation are not necessarily informative since one can't infer the
distribution from just these two numbers. The properties described above are specific to the
normal. For example, the normal distribution does not seem to be a good approximation
for the direct compensation for 199 United States CEOs in the year 2000.

```
data(exec.pay,package="UsingR")
mypar(1,2)
hist(exec.pay)
qqnorm(exec.pay)
qqline(exec.pay)
```

In addition to qq-plots, a practical summary of data is to compute 3 percentiles: 25th,
50th (the median) and the 75th. A boxplot shows these 3 values along with a range of the
points within median ± 1.5 (75th percentile - 25th percentile). Values outside this range
are shown as points and sometimes referred to as *outliers*.

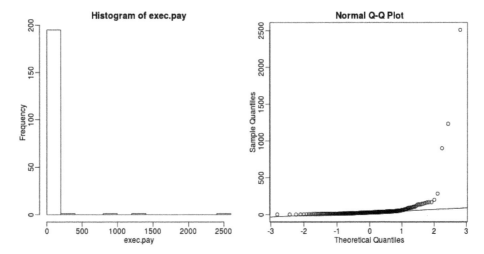

FIGURE 3.5
Histogram and QQ-plot of executive pay.

```
boxplot(exec.pay, ylab="10,000s of dollars", ylim=c(0,400))
```

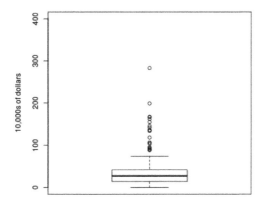

FIGURE 3.6
Simple boxplot of executive pay.

Here we show just one boxplot. However, one of the great benefits of boxplots is that we could easily show many distributions in one plot, by lining them up, side by side. We will see several examples of this throughout the book.

3.3   Scatterplots and Correlation

The methods described above relate to *univariate* variables. In the biomedical sciences, it is common to be interested in the relationship between two or more variables. A classic example is the father/son height data used by Francis Galton[1] to understand heredity. If we were to summarize these data, we could use the two averages and two standard deviations since both distributions are well approximated by the normal distribution. This summary, however, fails to describe an important characteristic of the data.

```
data(father.son,package="UsingR")
x=father.son$fheight
y=father.son$sheight
plot(x,y, xlab="Father's height in inches",
    ylab="Son's height in inches",
    main=paste("correlation =",signif(cor(x,y),2)))
```

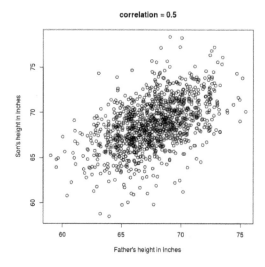

FIGURE 3.7
Heights of father and son pairs plotted against each other.

The scatter plot shows a general trend: the taller the father, the taller the son. A summary of this trend is the correlation coefficient, which in this case is 0.5. We will motivate this statistic by trying to predict the son's height using the father's height.

3.4   Stratification

Suppose we are asked to guess the height of randomly selected sons. The average height, 68.7 inches, is the value with the highest proportion (see histogram) and would be our prediction. But what if we are told that the father is 72 inches tall, do we still guess 68.7?

---
[1]https://en.wikipedia.org/wiki/Francis_Galton

The father is taller than average. Specifically, he is 1.75 standard deviations taller than the average father. So should we predict that the son is also 1.75 standard deviations taller? It turns out that this would be an overestimate. To see this, we look at all the sons with fathers who are about 72 inches. We do this by *stratifying* the father heights.

```
groups <- split(y,round(x))
boxplot(groups)
```

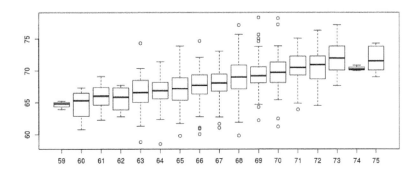

FIGURE 3.8
Boxplot of son heights stratified by father heights.

```
print(mean(y[ round(x) == 72]))
```

```
## [1] 70.67719
```

Stratification followed by boxplots lets us see the distribution of each group. The average height of sons with fathers that are 72 inches tall is 70.7 inches. We also see that the *medians* of the strata appear to follow a straight line (remember the middle line in the boxplot shows the median, not the mean). This line is similar to the *regression line*, with a slope that is related to the correlation, as we will learn below.

3.5 Bivariate Normal Distribution

Correlation is a widely used summary statistic in the life sciences. However, it is often misused or misinterpreted. To properly interpret correlation we actually have to understand the bivariate normal distribution.

A pair of random variables (X, Y) is considered to be approximated by bivariate normal when the proportion of values below, for example a and b, is approximated by this expression:

$$\Pr(X < a, Y < b) = \int_{-\infty}^{a} \int_{-\infty}^{b} \frac{1}{2\pi\sigma_x\sigma_y\sqrt{1-\rho^2}}$$
$$\exp\left(\frac{1}{2(1-\rho^2)}\left[\left(\frac{x-\mu_x}{\sigma_x}\right)^2 - 2\rho\left(\frac{x-\mu_x}{\sigma_x}\right)\left(\frac{y-\mu_y}{\sigma_y}\right) + \left(\frac{y-\mu_y}{\sigma_y}\right)^2\right]\right)$$

This may seem like a rather complicated equation, but the concept behind it is rather intuitive. An alternative definition is the following: fix a value x and look at all the pairs (X, Y) for which $X = x$. Generally, in statistics we call this exercise *conditioning*. We are conditioning Y on X. If a pair of random variables is approximated by a bivariate normal distribution, then the distribution of Y conditioned on $X = x$ is approximated with a normal distribution, no matter what x we choose. Let's see if this holds with our height data. We show 4 different strata:

```
groups <- split(y,round(x))
mypar(2,2)
for(i in c(5,8,11,14)){
  qqnorm(groups[[i]],main=paste0("X=",names(groups)[i]," strata"),
         ylim=range(y),xlim=c(-2.5,2.5))
  qqline(groups[[i]])
}
```

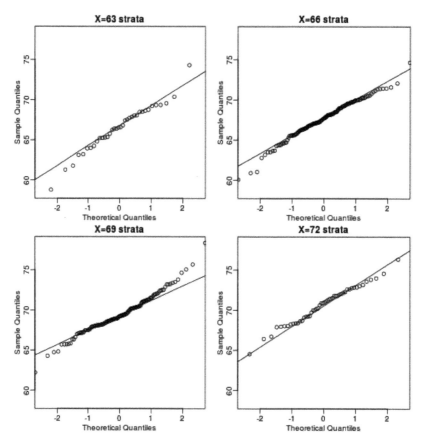

FIGURE 3.9
qqplots of son heights for four strata defined by father heights.

Now we come back to defining correlation. Mathematical statistics tells us that when two variables follow a bivariate normal distribution, then for any given value of x, the average of the Y in pairs for which $X = x$ is:

$$\mu_Y + \rho \frac{X - \mu_X}{\sigma_X} \sigma_Y$$

Note that this is a line with slope $\rho \frac{\sigma_Y}{\sigma_X}$. This is referred to as the *regression line.* If the SDs are the same, then the slope of the regression line is the correlation ρ. Therefore, if we standardize X and Y, the correlation is the slope of the regression line.

Another way to see this is to form a prediction $\hat{Y}$: for every SD away from the mean in x, we predict ρ SDs away for Y:

$$\frac{\hat{Y} - \mu_Y}{\sigma_Y} = \rho \frac{x - \mu_X}{\sigma_X}$$

If there is perfect correlation, we predict the same number of SDs. If there is 0 correlation, then we don't use x at all. For values between 0 and 1, the prediction is somewhere in between. For negative values, we simply predict in the opposite direction.

To confirm that the above approximations hold in this case, let's compare the mean of each strata to the identity line and the regression line:

```
x=( x-mean(x) )/sd(x)
y=( y-mean(y) )/sd(y)
means=tapply(y, round(x*4)/4, mean)
fatherheights=as.numeric(names(means))
mypar(1,1)
plot(fatherheights, means, ylab="average of strata of son heights", ylim=range(fatherheights))
abline(0, cor(x,y))
```

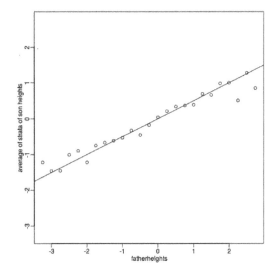

FIGURE 3.10
Average son height of each strata plotted against father heights defining the strata

Variance explained The standard deviation of the *conditional* distribution described above is:

$$\sqrt{1 - \rho^2} \sigma_Y$$

This is where statements like X explains $\rho^2 \times 100$ % of the variation in Y: the variance of Y is σ^2 and, once we condition, it goes down to $(1 - \rho^2)\sigma_Y^2$. It is important to remember that the "variance explained" statement only makes sense when the data is approximated by a bivariate normal distribution.

_____

3.6   Plots to Avoid

This section is based on a talk by Karl W. Broman[2] titled "How to Display Data Badly," in which he described how the default plots offered by Microsoft Excel "obscure your data and annoy your readers" (here[3] is a link to a collection of Karl Broman's talks). His lecture was inspired by the 1984 paper by H. Wainer: How to display data badly. American Statistician 38(2): 137–147. Dr. Wainer was the first to elucidate the principles of the bad display of data. However, according to Karl Broman, "The now widespread use of Microsoft Excel has resulted in remarkable advances in the field." Here we show examples of "bad plots" and how to improve them in R.

General principles The aim of good data graphics is to display data accurately and clearly. According to Karl Broman, some rules for displaying data *badly* are:

- Display as little information as possible.
- Obscure what you do show (with chart junk).
- Use pseudo-3D and color gratuitously.
- Make a pie chart (preferably in color and 3D).
- Use a poorly chosen scale.
- Ignore significant figures.

Pie charts Let's say we want to report the results from a poll asking about browser preference (taken in August 2013). The standard way of displaying these is with a pie chart:

```
pie(browsers,main="Browser Usage (August 2013)")
```

Nonetheless, as stated by the help file for the **pie** function:

"Pie charts are a very bad way of displaying information. The eye is good at judging linear measures and bad at judging relative areas. A bar chart or dot chart is a preferable way of displaying this type of data."

To see this, look at the figure above and try to determine the percentages just from looking at the plot. Unless the percentages are close to 25%, 50% or 75%, this is not so easy. Simply showing the numbers is not only clear, but also saves on printing costs.

```
browsers
```

```
##   Opera  Safari Firefox     IE  Chrome
##       1       9      20     26      44
```

If you do want to plot them, then a barplot is appropriate. Here we add horizontal lines at every multiple of 10 and then redraw the bars:

_____

[2]http://kbroman.org/
[3]http://kbroman.org/pages/talks.html

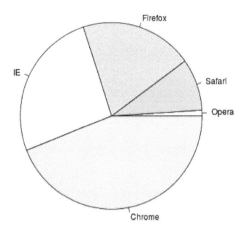

FIGURE 3.11
Pie chart of browser usage.

```
barplot(browsers, main="Browser Usage (August 2013)", ylim=c(0,55))
abline(h=1:5 * 10)
barplot(browsers, add=TRUE)
```

Notice that we can now pretty easily determine the percentages by following a horizontal line to the x-axis. Do avoid a 3D version since it obfuscates the plot, making it more difficult to find the percentages by eye.

Even worse than pie charts are donut plots.

The reason is that by removing the center, we remove one of the visual cues for determining the different areas: the angles. There is no reason to ever use a donut plot to display data.

Barplots as data summaries While barplots are useful for showing percentages, they are incorrectly used to display data from two groups being compared. Specifically, barplots are created with height equal to the group means; an antenna is added at the top to represent standard errors. This plot is simply showing two numbers per group and the plot adds nothing:

Much more informative is to summarize with a boxplot. If the number of points is small enough, we might as well add them to the plot. When the number of points is too large for us to see them, just showing a boxplot is preferable. We can even set **range=0** in **boxplot** to avoid drawing many outliers when the data is in the range of millions.

Let's recreate these barplots as boxplots. First let's download the data:

```
library(downloader)
filename <- "fig1.RData"
url <- "https://github.com/kbroman/Talk_Graphs/raw/master/R/fig1.RData"
if (!file.exists(filename)) download(url,filename)
load(filename)
```

Now we can simply show the points and make simple boxplots:

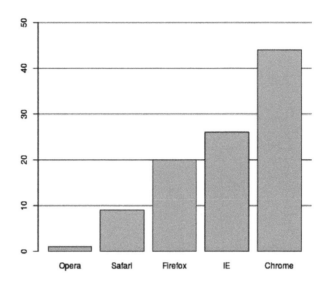

FIGURE 3.12
Barplot of browser usage.

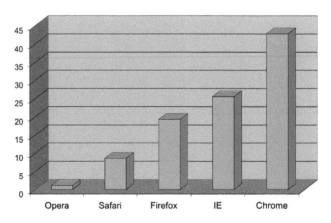

FIGURE 3.13
3D version.

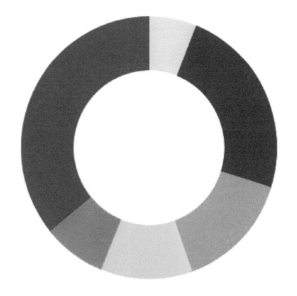

FIGURE 3.14
Donut plot.

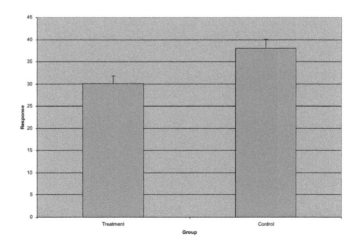

FIGURE 3.15
Bad bar plots.

```
library(rafalib)
mypar()
dat <- list(Treatment=x,Control=y)
boxplot(dat,xlab="Group",ylab="Response",cex=0)
stripchart(dat,vertical=TRUE,method="jitter",pch=16,add=TRUE,col=1)
```

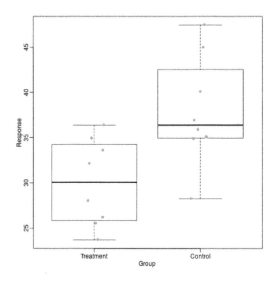

FIGURE 3.16
Treatment data and control data shown with a boxplot.

Notice how much more we see here: the center, spread, range, and the points themselves. In the barplot, we only see the mean and the SE, and the SE has more to do with sample size than with the spread of the data.

This problem is magnified when our data has outliers or very large tails. In the plot below, there appears to be very large and consistent differences between the two groups:

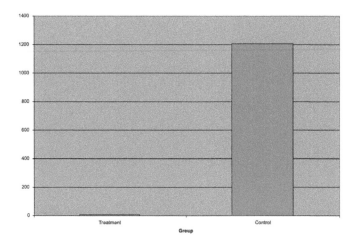

FIGURE 3.17
Bar plots with outliers.

However, a quick look at the data demonstrates that this difference is mostly driven by just two points. A version showing the data in the log-scale is much more informative.

Start by downloading data:

```
library(downloader)
url <- "https://github.com/kbroman/Talk_Graphs/raw/master/R/fig3.RData"
filename <- "fig3.RData"
if (!file.exists(filename)) download(url, filename)
load(filename)
```

Now we can show data and boxplots in original scale and log-scale.

```
library(rafalib)
mypar(1,2)
dat <- list(Treatment=x,Control=y)

boxplot(dat,xlab="Group",ylab="Response",cex=0)
stripchart(dat,vertical=TRUE,method="jitter",pch=16,add=TRUE,col=1)

boxplot(dat,xlab="Group",ylab="Response",log="y",cex=0)
stripchart(dat,vertical=TRUE,method="jitter",pch=16,add=TRUE,col=1)
```

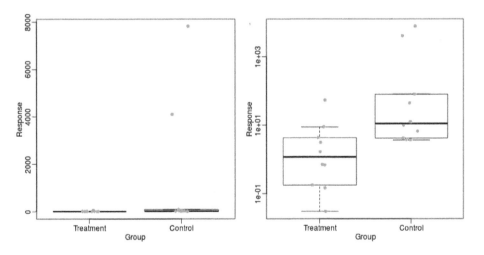

FIGURE 3.18
Data and boxplots for original data (left) and in log scale (right).

Show the scatter plot The purpose of many statistical analyses is to determine relationships between two variables. Sample correlations are typically reported and sometimes plots are displayed to show this. However, showing just the regression line is one way to display your data badly since it hides the scatter. Surprisingly, plots such as the following are commonly seen.

Again start by loading data:

```
url <- "https://github.com/kbroman/Talk_Graphs/raw/master/R/fig4.RData"
filename <- "fig4.RData"
if (!file.exists(filename)) download(url, filename)
load(filename)
```

```
mypar(1,2)
plot(x,y,lwd=2,type="n")
fit <- lm(y~x)
abline(fit$coef,lwd=2)
b <- round(fit$coef,4)
text(78, 200, paste("y =", b[1], "+", b[2], "x"), adj=c(0,0.5))
rho <- round(cor(x,y),4)
text(78, 187,expression(paste(rho," = 0.8567")),adj=c(0,0.5))

plot(x,y,lwd=2)
fit <- lm(y~x)
abline(fit$coef,lwd=2)
```

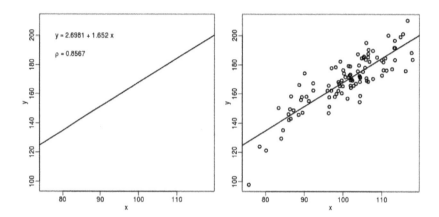

FIGURE 3.19

The plot on the left shows a regression line that was fitted to the data shown on the right. It is much more informative to show all the data.

When there are large amounts of points, the scatter can be shown by binning in two dimensions and coloring the bins by the number of points in the bin. An example of this is the hexbin function in the hexbin package[4].

High correlation does not imply replication When new technologies or laboratory techniques are introduced, we are often shown scatter plots and correlations from replicated samples. High correlations are used to demonstrate that the new technique is reproducible. Correlation, however, can be very misleading. Below is a scatter plot showing data from replicated samples run on a high throughput technology. This technology outputs 12,626 simultaneous measurements.

In the plot on the left, we see the original data which shows very high correlation. Yet the data follows a distribution with very fat tails. Furthermore, 95% of the data is below the green line. The plot on the right is in the log scale:

Note that we do not show the code here as it is rather complex but we explain how to make MA plots in a later chapter.

Although the correlation is reduced in the log-scale, it is very close to 1 in both cases. Does this mean these data are reproduced? To examine how well the second vector repro-

---

[4]https://cran.r-project.org/package=hexbin

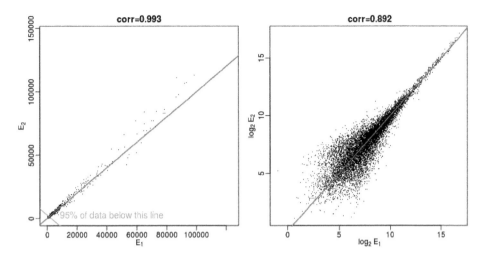

FIGURE 3.20
Gene expression data from two replicated samples. Left is in original scale and right is in log scale.

duces the first, we need to study the differences. We therefore should plot that instead. In this plot, we plot the difference (in the log scale) versus the average:

These are referred to as Bland-Altman plots, or *MA plots* in the genomics literature, and we will talk more about them later. "MA" stands for "minus" and "average" because in this plot, the y-axis is the difference between two samples on the log scale (the log ratio is the difference of the logs), and the x-axis is the average of the samples on the log scale. In this plot, we see that the typical difference in the log (base 2) scale between two replicated measures is about 1. This means that when measurements should be the same, we will, on average, observe 2 fold difference. We can now compare this variability to the differences we want to detect and decide if this technology is precise enough for our purposes.

Barplots for paired data A common task in data analysis is the comparison of two groups. When the dataset is small and data are paired, such as the outcomes before and after a treatment, two-color barplots are unfortunately often used to display the results.

There are better ways of showing these data to illustrate that there is an increase after treatment. One is to simply make a scatter plot, which shows that most points are above the identity line. Another alternative is to plot the differences against the before values.

```
set.seed(12201970)
before <- runif(6, 5, 8)
after <- rnorm(6, before*1.05, 2)
li <- range(c(before, after))
ymx <- max(abs(after-before))

mypar(1,2)
plot(before, after, xlab="Before", ylab="After",
     ylim=li, xlim=li)
abline(0,1, lty=2, col=1)

plot(before, after-before, xlab="Before", ylim=c(-ymx, ymx),
```

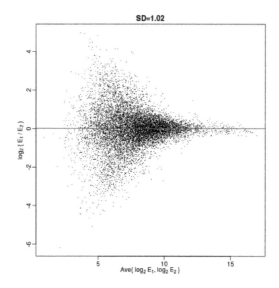

FIGURE 3.21
MA plot of the same data shown above shows that data is not replicated very well despite a high correlation.

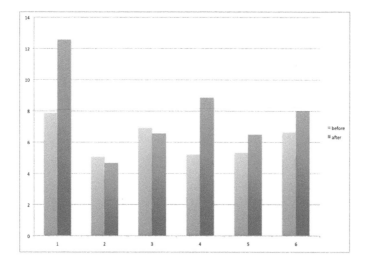

FIGURE 3.22
Barplot for two variables.

```
    ylab="Change (After - Before)", lwd=2)
abline(h=0, lty=2, col=1)
```

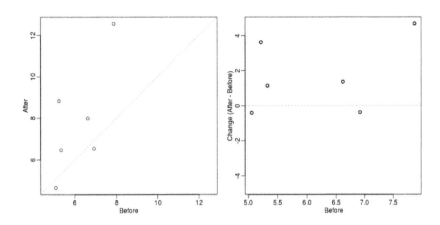

FIGURE 3.23
For two variables a scatter plot or a 'rotated' plot similar to an MA plot is much more informative.

Line plots are not a bad choice, although I find them harder to follow than the previous two. Boxplots show you the increase, but lose the paired information.

```
z <- rep(c(0,1), rep(6,2))
mypar(1,2)
plot(z, c(before, after),
    xaxt="n", ylab="Response",
    xlab="", xlim=c(-0.5, 1.5))
axis(side=1, at=c(0,1), c("Before","After"))
segments(rep(0,6), before, rep(1,6), after, col=1)

boxplot(before,after,names=c("Before","After"),ylab="Response")
```

Gratuitous 3D The figure below shows three curves. Pseudo 3D is used, but it is not clear why. Maybe to separate the three curves? Notice how difficult it is to determine the values of the curves at any given point:
  This plot can be made better by simply using color to distinguish the three lines:

```
##First read data
url <- "https://github.com/kbroman/Talk_Graphs/raw/master/R/fig8dat.csv"
x <- read.csv(url)

##Now make alternative plot
plot(x[,1],x[,2],xlab="log Dose",ylab="Proportion survived",ylim=c(0,1),
    type="l",lwd=2,col=1)
lines(x[,1],x[,3],lwd=2,col=2)
lines(x[,1],x[,4],lwd=2,col=3)
legend(1,0.4,c("Drug A","Drug B","Drug C"),lwd=2, col=1:3)
```

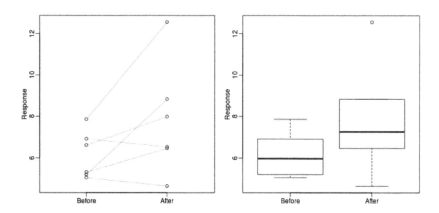

FIGURE 3.24
Another alternative is a line plot. If we don't care about pairings, then the boxplot is appropriate.

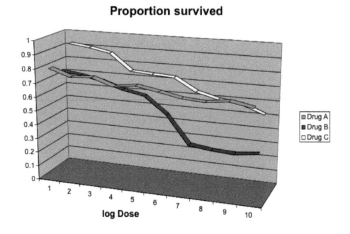

FIGURE 3.25
Gratuitous 3-D.

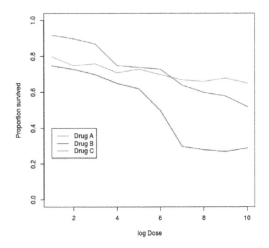

FIGURE 3.26
This plot demonstrates that using color is more than enough to distinguish the three lines.

Ignoring important factors In this example, we generate data with a simulation. We are studying a dose-response relationship between two groups: treatment and control. We have three groups of measurements for both control and treatment. Comparing treatment and control using the common barplot.

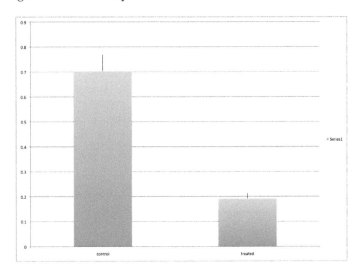

FIGURE 3.27
Ignoring important factors.

Instead, we should show each curve. We can use color to distinguish treatment and control, and dashed and solid lines to distinguish the original data from the mean of the three groups.

```
plot(x, y1, ylim=c(0,1), type="n", xlab="Dose", ylab="Response")
for(i in 1:3) lines(x, y[,i], col=1, lwd=1, lty=2)
```

```
for(i in 1:3) lines(x, z[,i], col=2, lwd=1, lty=2)
lines(x, ym, col=1, lwd=2)
lines(x, zm, col=2, lwd=2)
legend("bottomleft", lwd=2, col=c(1, 2), c("Control", "Treated"))
```

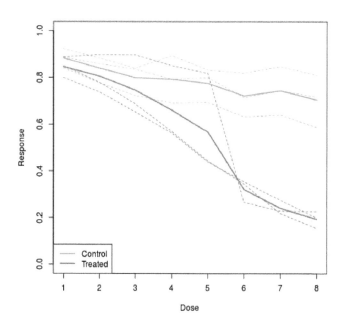

FIGURE 3.28
Because dose is an important factor, we show it in this plot.

Too many significant digits By default, statistical software like R returns many significant digits. This does not mean we should report them. Cutting and pasting directly from R is a bad idea since you might end up showing a table, such as the one below, comparing the heights of basketball players:

```
heights <- cbind(rnorm(8,73,3),rnorm(8,73,3),rnorm(8,80,3),
                 rnorm(8,78,3),rnorm(8,78,3))
colnames(heights)<-c("SG","PG","C","PF","SF")
rownames(heights)<- paste("team",1:8)
heights
```

```
##               SG       PG        C       PF       SF
## team 1 76.39843 76.21026 81.68291 75.32815 77.18792
## team 2 74.14399 71.10380 80.29749 81.58405 73.01144
## team 3 71.51120 69.02173 85.80092 80.08623 72.80317
## team 4 78.71579 72.80641 81.33673 76.30461 82.93404
## team 5 73.42427 73.27942 79.20283 79.71137 80.30497
## team 6 72.93721 71.81364 77.35770 81.69410 80.39703
## team 7 68.37715 73.01345 79.10755 71.24982 77.19851
## team 8 73.77538 75.59278 82.99395 75.57702 87.68162
```

We are reporting precision up to 0.00001 inches. Do you know of a tape measure with that much precision? This can be easily remedied:

```
round(heights,1)
```

```
##            SG   PG    C   PF   SF
## team 1 76.4 76.2 81.7 75.3 77.2
## team 2 74.1 71.1 80.3 81.6 73.0
## team 3 71.5 69.0 85.8 80.1 72.8
## team 4 78.7 72.8 81.3 76.3 82.9
## team 5 73.4 73.3 79.2 79.7 80.3
## team 6 72.9 71.8 77.4 81.7 80.4
## team 7 68.4 73.0 79.1 71.2 77.2
## team 8 73.8 75.6 83.0 75.6 87.7
```

Displaying data well In general, you should follow these principles:

- Be accurate and clear.
- Let the data speak.
- Show as much information as possible, taking care not to obscure the message.
- Science not sales: avoid unnecessary frills (especially gratuitous 3D).
- In tables, every digit should be meaningful. Don't drop ending 0's.

Some further reading:

- ER Tufte (1983) The visual display of quantitative information. Graphics Press.
- ER Tufte (1990) Envisioning information. Graphics Press.
- ER Tufte (1997) Visual explanations. Graphics Press.
- WS Cleveland (1993) Visualizing data. Hobart Press.
- WS Cleveland (1994) The elements of graphing data. CRC Press.
- A Gelman, C Pasarica, R Dodhia (2002) Let's practice what we preach: Turning tables into graphs. The American Statistician 56:121-130.
- NB Robbins (2004) Creating more effective graphs. Wiley.
- Nature Methods columns[5]

3.7 Misunderstanding Correlation (Advanced)

The use of correlation to summarize reproducibility has become widespread in, for example, genomics. Despite its English language definition, mathematically, correlation is not necessarily informative with regards to reproducibility. Here we briefly describe three major problems.

The most egregious related mistake is to compute correlations of data that are not approximated by bivariate normal data. As described above, averages, standard deviations and correlations are popular summary statistics for two-dimensional data because, for the bivariate normal distribution, these five parameters fully describe the distribution. However, there are many examples of data that are not well approximated by bivariate normal data.

---

[5]http://bang.clearscience.info/?p=546

Raw gene expression data, for example, tends to have a distribution with a very fat right tail.

The standard way to quantify reproducibility between two sets of replicated measurements, say $x_1, \ldots, x_n$ and $y_1, \ldots, y_n$, is simply to compute the distance between them:

$$\sqrt{\sum_{i=1}^{n} d_i^2} \text{ with } d_i = x_i - y_i$$

This metric decreases as reproducibility improves and it is 0 when the reproducibility is perfect. Another advantage of this metric is that if we divide the sum by N, we can interpret the resulting quantity as the standard deviation of the $d_1, \ldots, d_N$ if we assume the d average out to 0. If the d can be considered residuals, then this quantity is equivalent to the root mean squared error (RMSE), a summary statistic that has been around for over a century. Furthermore, this quantity will have the same units as our measurements resulting in a more interpretable metric.

Another limitation of the correlation is that it does not detect cases that are not reproducible due to average changes. The distance metric does detect these differences. We can rewrite:

$$\frac{1}{n} \sum_{i=1}^{n} (x_i - y_i)^2 = \frac{1}{n} \sum_{i=1}^{n} [(x_i - \mu_x) - (y_i - \mu_y) + (\mu_x - \mu_y)]^2$$

with μ_x and μ_y the average of each list. Then we have:

$$\frac{1}{n} \sum_{i=1}^{n} (x_i - y_i)^2 = \frac{1}{n} \sum_{i=1}^{n} (x_i - \mu_x)^2 + \frac{1}{n} \sum_{i=1}^{n} (y_i - \mu_y)^2 + (\mu_x - \mu_y)^2 + \frac{1}{n} \sum_{i=1}^{n} (x_i - \mu_x)(y_i - \mu_y)$$

For simplicity, if we assume that the variance of both lists is 1, then this reduces to:

$$\frac{1}{n} \sum_{i=1}^{n} (x_i - y_i)^2 = 2 + (\mu_x - \mu_y)^2 - 2\rho$$

with ρ the correlation. So we see the direct relationship between distance and correlation. However, an important difference is that the distance contains the term

$$(\mu_x - \mu_y)^2$$

and, therefore, it can detect cases that are not reproducible due to large average changes.

Yet another reason correlation is not an optimal metric for reproducibility is the lack of units. To see this, we use a formula that relates the correlation of a variable with that variable, plus what is interpreted here as deviation: x and $y = x + d$. The larger the variance of d, the less $x + d$ reproduces x. Here the distance metric would depend only on the variance of d and would summarize reproducibility. However, correlation depends on the variance of x as well. If d is independent of x, then

$$\text{cor}(x, y) = \frac{1}{\sqrt{1 + \text{var}(d)/\text{var}(x)}}$$

This suggests that correlations near 1 do not necessarily imply reproducibility. Specifically, irrespective of the variance of d, we can make the correlation arbitrarily close to 1 by increasing the variance of x. {pagebreak}

3.8   Exercises

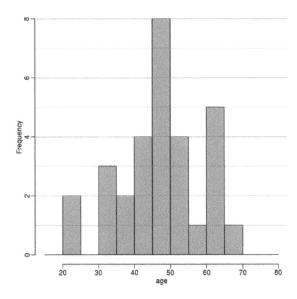

FIGURE 3.29

1. Given the above histogram, how many people are between the ages of 35 and 45?

2. The `InsectSprays` dataset is included in R. The dataset reports the counts of insects in agricultural experimental units treated with different insecticides. Make a boxplot and determine which insecticide appears to be most effective (has the lowest median).

3. Download and load this[6] dataset into R. Use exploratory data analysis tools to determine which two columns are different from the rest. Which column has positive skew (a long tail to the right)?

4. Which column has negative skew (a long tail to the left)?

5. Let's consider a random sample of finishers from the New York City Marathon in 2002. This dataset can be found in the UsingR package. Load the library and then load the `nym.2002` dataset.

```
library(dplyr)
data(nym.2002, package="UsingR")
```

Use boxplots and histograms to compare the finishing times of males and females. Which of the following best describes the difference?

(a) Males and females have the same distribution.

(b) Most males are faster than most women.

---

[6]http://courses.edx.org/c4x/HarvardX/PH525.1x/asset/skew.RData

(c) Males and females have similar right skewed distributions, with the former 20 minutes shifted to the left.

(d) Both distributions are normally distributed with a difference in mean of about 30 minutes.

6. Use `dplyr` to create two new data frames: `males` and `females`, with the data for each gender. For males, what is the Pearson correlation between age and time to finish?

7. For females, what is the Pearson correlation between age and time to finish?

8. If we interpret these correlations without visualizing the data, we would conclude that the older we get, the slower we run marathons, regardless of gender. Look at scatterplots and boxplots of times stratified by age groups (20-25, 25-30, etc.). After examining the data, what is a more reasonable conclusion?

(a) Finish times are constant up until about our 40s, then we get slower.

(b) On average, finish times go up by about 7 minutes every five years.

(c) The optimal age to run a marathon is 20-25.

(d) Coding errors never happen: a five-year-old boy completed the 2012 NY city marathon.

9. When is it appropriate to use pie charts or donut charts?

(a) When you are hungry.

(b) To compare percentages.

(c) To compare values that add up to 100%.

(d) Never.

10. The use of pseudo-3D plots in the literature mostly adds:

(a) Pizzazz.

(b) The ability to see three dimensional data.

(c) Ability to discover.

(d) Confusion.

3.9 Robust Summaries

The normal approximation is often useful when analyzing life sciences data. However, due to the complexity of the measurement devices, it is also common to mistakenly observe data points generated by an undesired process. For example, a defect on a scanner can produce a handful of very high intensities or a PCR bias can lead to a fragment appearing much more often than others. We therefore may have situations that are approximated by, for example, 99 data points from a standard normal distribution and one large number.

```
set.seed(1)
x=c(rnorm(100,0,1)) ##real distribution
x[23] <- 100 ##mistake made in 23th measurement
boxplot(x)
```

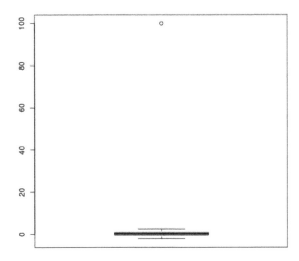

FIGURE 3.30
Normally distributed data with one point that is very large due to a mistake.

In statistics we refer to these type of points as *outliers*. A small number of outliers can throw off an entire analysis. For example, notice how the following one point results in the sample mean and sample variance being very far from the 0 and 1 respectively.

```
cat("The average is",mean(x),"and the SD is",sd(x))
```

```
## The average is 1.108142 and the SD is 10.02938
```

The median The median, defined as the point having half the data larger and half the data smaller, is a summary statistic that is *robust* to outliers. Note how much closer the median is to 0, the center of our actual distribution:

```
median(x)
```

```
## [1] 0.1684483
```

The median absolute deviation The median absolute deviation (MAD) is a robust summary for the standard deviation. It is defined by computing the differences between each point and the median, and then taking the median of their absolute values:

$$1.4826 \text{median}|X_i - \text{median}(X_i)|$$

The number 1.4826 is a scaling factor such that the MAD is an unbiased estimate of the standard deviation. Notice how much closer we are to 1 with the MAD:

```
mad(x)
```

```
## [1] 0.8857141
```

Spearman correlation Earlier we saw that the correlation is also sensitive to outliers. Here we construct an independent list of numbers, but for which a similar mistake was made for the same entry:

```
set.seed(1)
x=c(rnorm(100,0,1)) ##real distribution
x[23] <- 100 ##mistake made in 23th measurement
y=c(rnorm(100,0,1)) ##real distribution
y[23] <- 84 ##similar mistake made in 23th measurement
library(rafalib)
mypar()
plot(x,y,main=paste0("correlation=",round(cor(x,y),3)),
     pch=21,bg=1,xlim=c(-3,100),ylim=c(-3,100))
abline(0,1)
```

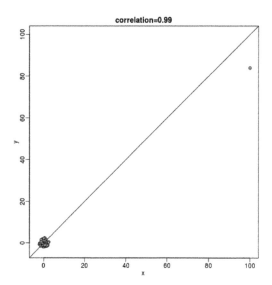

FIGURE 3.31
Scatterplot showing bivariate normal data with one signal outlier resulting in large values in both dimensions.

The Spearman correlation follows the general idea of median and MAD, that of using quantiles. The idea is simple: we convert each dataset to ranks and then compute correlation:

```
mypar(1,2)
plot(x,y,main=paste0("correlation=",round(cor(x,y),3)),pch=21,bg=1,xlim=c(-3,100),ylim=c(-3,100))
plot(rank(x),rank(y), main=paste0("correlation=",round(cor(x,y,method="spearman"),3)),
     pch=21,bg=1,xlim=c(-3,100),ylim=c(-3,100))
abline(0,1)
```

So if these statistics are robust to outliers, why would we ever use the non-robust version? In general, if we know there are outliers, then median and MAD are recommended over the mean and standard deviation counterparts. However, there are examples in which robust statistics are less powerful than the non-robust versions.

We also note that there is a large statistical literature on robust statistics that go far beyond the median and the MAD.

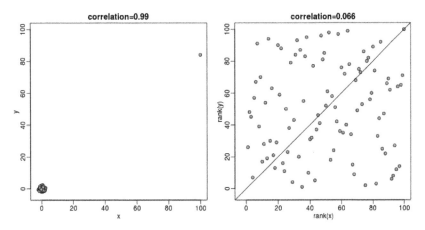

FIGURE 3.32
Scatterplot of original data (left) and ranks (right). Spearman correlation reduces the influence of outliers by considering the ranks instead of original data.

Symmetry of log ratios Ratios are not symmetric. To see this, we will start by simulating the ratio of two positive random numbers, which will represent the expression of genes in two samples:

```
x <- 2^(rnorm(100))
y <- 2^(rnorm(100))
ratios <- x / y
```

Reporting ratios or fold changes are common in the life sciences. Suppose you are studying ratio data showing, say, gene expression before and after treatment. You are given ratio data so values larger than 1 imply gene expression was higher after the treatment. If the treatment has no effect, we should see as many values below 1 as above 1. A histogram seems to suggest that the treatment does in fact have an effect:

```
mypar(1,2)
hist(ratios)

logratios <- log2(ratios)
hist(logratios)
```

The problem here is that ratios are not symmetrical around 1. For example, 1/32 is much closer to 1 than 32/1. Using the log takes care of this problem. The log of ratios are of course symmetric around 0 because:

$$\log(x/y) = \log(x) - \log(y) = -(\log(y) - \log(x)) = \log(y/x)$$

As demonstrated by these simple plots:
In the life sciences, the log transformation is also commonly used because (multiplicative) fold changes are the most widely used quantification of interest. Note that a fold change of 100 can be a ratio of 100/1 or 1/100. However, 1/100 is much closer to 1 (no fold change) than 100: ratios are not symmetric about 1.

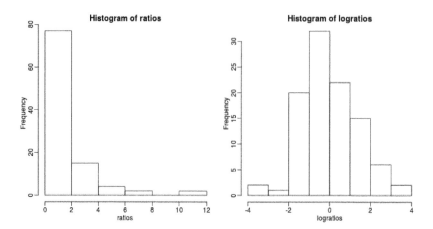

FIGURE 3.33
Histogram of original (left) and log (right) ratios.

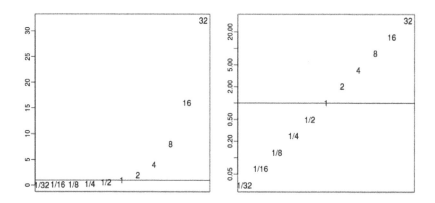

FIGURE 3.34
Histogram of original (left) and log (right) powers of 2 seen as ratios.

3.10   Wilcoxon Rank Sum Test

We learned how the sample mean and SD are susceptible to outliers. The t-test is based on these measures and is susceptible as well. The Wilcoxon rank test (equivalent to the Mann-Whitney test) provides an alternative. In the code below, we perform a t-test on data for which the null is true. However, we change one sum observation by mistake in each sample and the values incorrectly entered are different. Here we see that the t-test results in a small p-value, while the Wilcoxon test does not:

```
set.seed(779) ##779 picked for illustration purposes
N=25
x<- rnorm(N,0,1)
y<- rnorm(N,0,1)
```

Create outliers:

```
x[1] <- 5
x[2] <- 7
cat("t-test pval:",t.test(x,y)$p.value)
```

t-test pval: 0.04439948

```
cat("Wilcox test pval:",wilcox.test(x,y)$p.value)
```

Wilcox test pval: 0.1310212

The basic idea is to 1) combine all the data, 2) turn the values into ranks, 3) separate them back into their groups, and 4) compute the sum or average rank and perform a test.

```
library(rafalib)
mypar(1,2)

stripchart(list(x,y),vertical=TRUE,ylim=c(-7,7),ylab="Observations",pch=21,bg=1)
abline(h=0)

xrank<-rank(c(x,y))[seq(along=x)]
yrank<-rank(c(x,y))[-seq(along=y)]

stripchart(list(xrank,yrank),vertical=TRUE,ylab="Ranks",pch=21,bg=1,cex=1.25)

ws <- sapply(x,function(z) rank(c(z,y))[1]-1)
text( rep(1.05,length(ws)), xrank, ws, cex=0.8)

W <-sum(ws)
```

W is the sum of the ranks for the first group relative to the second group. We can compute an exact p-value for W based on combinatorics. We can also use the CLT since statistical theory tells us that this W is approximated by the normal distribution. We can construct a z-score as follows:

```
n1<-length(x);n2<-length(y)
Z <- (mean(ws)-n2/2)/ sqrt(n2*(n1+n2+1)/12/n1)
print(Z)
```

[1] 1.523124

Here the Z is not large enough to give us a p-value less than 0.05. These are part of the calculations performed by the R function `wilcox.test`.

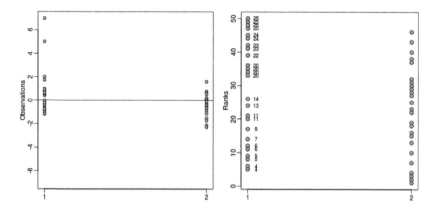

FIGURE 3.35
Data from two populations with two outliers. The left plot shows the original data and the right plot shows their ranks. The numbers are the w values

---

3.11   Exercises

We are going to explore the properties of robust statistics. We will use one of the datasets included in R, which contains weight of chicks in grams as they grow from day 0 to day 21. This dataset also splits up the chicks by different protein diets, which are coded from 1 to 4. We use this dataset to also show an important operation in R (not related to robust summaries): `reshape`.

This dataset is built into R and can be loaded with:

```
data(ChickWeight)
```

To begin, take a look at the weights of all observations over time and color the points to represent the Diet:

```
head(ChickWeight)
plot(ChickWeight$Time, ChickWeight$weight, col=ChickWeight$Diet)
```

First, notice that the rows here represent time points rather than individuals. To facilitate the comparison of weights at different time points and across the different chicks, we will reshape the data so that each row is a chick. In R we can do this with the `reshape` function:

```
chick = reshape(ChickWeight, idvar=c("Chick","Diet"), timevar="Time",
direction="wide")
```

The meaning of this line is: reshape the data from *long* to *wide*, where the columns Chick and Diet are the ID's and the column Time indicates different observations for each ID. Now examine the head of this dataset:

```
head(chick)
```

We also want to remove any chicks that have missing observations at any time points (NA for "not available"). The following line of code identifies these rows and then removes them:

```
chick = na.omit(chick)
```

1. Focus on the chick weights on day 4 (check the column names of 'chick' and note the numbers). How much does the average of chick weights at day 4 increase if we add an outlier measurement of 3000 grams? Specifically, what is the average weight of the day 4 chicks, including the outlier chick, divided by the average of the weight of the day 4 chicks without the outlier. Hint: use `c` to add a number to a vector.

2. In exercise 1, we saw how sensitive the mean is to outliers. Now let's see what happens when we use the median instead of the mean. Compute the same ratio, but now using median instead of mean. Specifically, what is the median weight of the day 4 chicks, including the outlier chick, divided by the median of the weight of the day 4 chicks without the outlier.

3. Now try the same thing with the sample standard deviation (the `sd` function in R). Add a chick with weight 3000 grams to the chick weights from day 4. How much does the standard deviation change? What's the standard deviation with the outlier chick divided by the standard deviation without the outlier chick?

4. Compare the result above to the median absolute deviation in R, which is calculated with the `mad` function. Note that the mad is unaffected by the addition of a single outlier. The `mad` function in R includes the scaling factor 1.4826, such that `mad` and `sd` are very similar for a sample from a normal distribution. What's the MAD with the outlier chick divided by the MAD without the outlier chick?

5. Our last question relates to how the Pearson correlation is affected by an outlier as compared to the Spearman correlation. The Pearson correlation between x and y is given in R by `cor(x,y)`. The Spearman correlation is given by `cor(x,y,method="spearman")`.

Plot the weights of chicks from day 4 and day 21. We can see that there is some general trend, with the lower weight chicks on day 4 having low weight again on day 21, and likewise for the high weight chicks.

Calculate the Pearson correlation of the weights of chicks from day 4 and day 21. Now calculate how much the Pearson correlation changes if we add a chick that weighs 3000 on day 4 and 3000 on day 21. Again, divide the Pearson correlation with the outlier chick over the Pearson correlation computed without the outliers.

6. Save the weights of the chicks on day 4 from diet 1 as a vector x. Save the weights of the chicks on day 4 from diet 4 as a vector y. Perform a t-test comparing x and y (in R the function `t.test(x,y)` will perform the test). Then perform a Wilcoxon test of x and y (in R the function `wilcox.test(x,y)` will perform the test). A warning will appear that an exact p-value cannot be calculated with ties, so an approximation is used, which is fine for our purposes.

Perform a t-test of x and y, after adding a single chick of weight 200 grams to x (the diet 1 chicks). What is the p-value from this test? The p-value of a test is available with the following code: `t.test(x,y)$p.value`

7. Do the same for the Wilcoxon test. The Wilcoxon test is robust to the outlier. In addition, it has fewer assumptions than the t-test on the distribution of the underlying data.

8. We will now investigate a possible downside to the Wilcoxon-Mann-Whitney test statistic. Using the following code to make three boxplots, showing the true Diet

1 vs 4 weights, and then two altered versions: one with an additional difference of 10 grams and one with an additional difference of 100 grams. Use the x and y as defined above, NOT the ones with the added outlier.

```
library(rafalib)
mypar(1,3)
boxplot(x,y)
boxplot(x,y+10)
boxplot(x,y+100)
```

What is the difference in t-test statistic (obtained by `t.test(x,y)$statistic`) between adding 10 and adding 100 to all the values in the group 'y'? Take the the t-test statistic with x and y+10 and subtract the t-test statistic with x and y+100. The value should be positive.

9. Examine the Wilcoxon test statistic for x and y+10 and for x and y+100. Because the Wilcoxon works on ranks, once the two groups show complete separation, that is, all points from group 'y' are above all points from group 'x', the statistic will not change, regardless of how large the difference grows. Likewise, the p-value has a minimum value, regardless of how far apart the groups are. This means that the Wilcoxon test can be considered less powerful than the t-test in certain contexts. In fact, for small sample sizes, the p-value can't be very small, even when the difference is very large. What is the p-value if we compare c(1,2,3) to c(4,5,6) using a Wilcoxon test?

10. What is the p-value if we compare c(1,2,3) to c(400,500,600) using a Wilcoxon test?

4

Matrix Algebra

In this book we try to minimize mathematical notation as much as possible. Furthermore, we avoid using calculus to motivate statistical concepts. However, Matrix Algebra (also referred to as Linear Algebra) and its mathematical notation greatly facilitates the exposition of the advanced data analysis techniques covered in the remainder of this book. We therefore dedicate a chapter of this book to introducing Matrix Algebra. We do this in the context of data analysis and using one of the main applications: Linear Models.

We will describe three examples from the life sciences: one from physics, one related to genetics, and one from a mouse experiment. They are very different, yet we end up using the same statistical technique: fitting linear models. Linear models are typically taught and described in the language of matrix algebra.

4.1   Motivating Examples

Falling objects   Imagine you are Galileo in the 16th century trying to describe the velocity of a falling object. An assistant climbs the Tower of Pisa and drops a ball, while several other assistants record the position at different times. Let's simulate some data using the equations we know today and adding some measurement error:

```
set.seed(1)
g <- 9.8 ##meters per second
n <- 25
tt <- seq(0,3.4,len=n) ##time in secs, note: we use tt because t is a base function
d <- 56.67  - 0.5*g*tt^2 + rnorm(n,sd=1) ##meters
```

The assistants hand the data to Galileo and this is what he sees:

```
mypar()
plot(tt,d,ylab="Distance in meters",xlab="Time in seconds")
```

He does not know the exact equation, but by looking at the plot above he deduces that the position should follow a parabola. So he models the data with:

$$Y_i = \beta_0 + \beta_1 x_i + \beta_2 x_i^2 + \varepsilon_i, i = 1, \ldots, n$$

With Y_i representing location, x_i representing the time, and ε_i accounting for measurement error. This is a linear model because it is a linear combination of known quantities (the x's) referred to as predictors or covariates and unknown parameters (the β's).

Father & son heights   Now imagine you are Francis Galton in the 19th century and you collect paired height data from fathers and sons. You suspect that height is inherited. Your data:

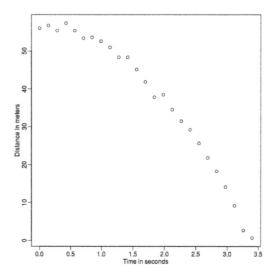

FIGURE 4.1
Simulated data for distance travelled versus time of falling object measured with error.

```
data(father.son,package="UsingR")
x=father.son$fheight
y=father.son$sheight
```

looks like this:

```
plot(x,y,xlab="Father's height",ylab="Son's height")
```

The sons' heights do seem to increase linearly with the fathers' heights. In this case, a model that describes the data is as follows:

$$Y_i = \beta_0 + \beta_1 x_i + \varepsilon_i, i = 1, \ldots, N$$

This is also a linear model with x_i and Y_i, the father and son heights respectively, for the i-th pair and ε_i a term to account for the extra variability. Here we think of the fathers' heights as the predictor and being fixed (not random) so we use lower case. Measurement error alone can't explain all the variability seen in ε_i. This makes sense as there are other variables not in the model, for example, mothers' heights, genetic randomness, and environmental factors.

Random samples from multiple populations Here we read-in mouse body weight data from mice that were fed two different diets: high fat and control (chow). We have a random sample of 12 mice for each. We are interested in determining if the diet has an effect on weight. Here is the data:

```
dat <- read.csv("femaleMiceWeights.csv")
mypar(1,1)
stripchart(Bodyweight~Diet,data=dat,vertical=TRUE,method="jitter",pch=1,main="Mice weights")
```

We want to estimate the difference in average weight between populations. We demonstrated how to do this using t-tests and confidence intervals, based on the difference in sample averages. We can obtain the same exact results using a linear model:

$$Y_i = \beta_0 + \beta_1 x_i + \varepsilon_i$$

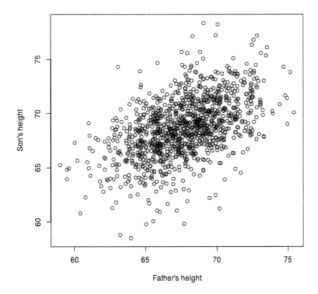

FIGURE 4.2
Galton's data. Son heights versus father heights.

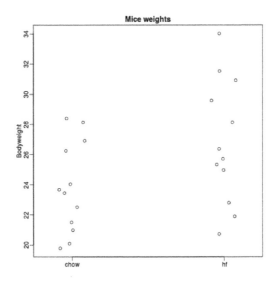

FIGURE 4.3
Mouse weights under two diets.

with β_0 the chow diet average weight, β_1 the difference between averages, $x_i = 1$ when mouse i gets the high fat (hf) diet, $x_i = 0$ when it gets the chow diet, and ε_i explains the differences between mice of the same population.

Linear models in general We have seen three very different examples in which linear models can be used. A general model that encompasses all of the above examples is the following:

$$Y_i = \beta_0 + \beta_1 x_{i,1} + \beta_2 x_{i,2} + \cdots + \beta_2 x_{i,p} + \varepsilon_i, i = 1, \ldots, n$$

$$Y_i = \beta_0 + \sum_{j=1}^{p} \beta_j x_{i,j} + \varepsilon_i, i = 1, \ldots, n$$

Note that we have a general number of predictors p. Matrix algebra provides a compact language and mathematical framework to compute and make derivations with any linear model that fits into the above framework.

Estimating parameters For the models above to be useful we have to estimate the unknown β s. In the first example, we want to describe a physical process for which we can't have unknown parameters. In the second example, we better understand inheritance by estimating how much, on average, the father's height affects the son's height. In the final example, we want to determine if there is in fact a difference: if $\beta_1 \neq 0$.

The standard approach in science is to find the values that minimize the distance of the fitted model to the data. The following is called the least squares (LS) equation and we will see it often in this chapter:

$$\sum_{i=1}^{n} \left\{ Y_i - \left(\beta_0 + \sum_{j=1}^{p} \beta_j x_{i,j} \right) \right\}^2$$

Once we find the minimum, we will call the values the least squares estimates (LSE) and denote them with $\hat{\beta}$. The quantity obtained when evaluating the least squares equation at the estimates is called the residual sum of squares (RSS). Since all these quantities depend on Y, *they are random variables*. The $\hat{\beta}$ s are random variables and we will eventually perform inference on them.

Falling object example revisited Thanks to my high school physics teacher, I know that the equation for the trajectory of a falling object is:

$$d = h_0 + v_0 t - 0.5 \times 9.8 t^2$$

with h_0 and v_0 the starting height and velocity respectively. The data we simulated above followed this equation and added measurement error to simulate n observations for dropping the ball ($v_0 = 0$) from the tower of Pisa ($h_0 = 56.67$). This is why we used this code to simulate data:

```
g <- 9.8 ##meters per second
n <- 25
tt <- seq(0,3.4,len=n) ##time in secs, t is a base function
f <- 56.67  - 0.5*g*tt^2
y <-  f + rnorm(n,sd=1)
```

Here is what the data looks like with the solid line representing the true trajectory:

```
plot(tt,y,ylab="Distance in meters",xlab="Time in seconds")
lines(tt,f,col=2)
```

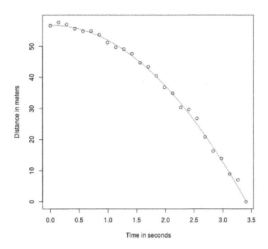

FIGURE 4.4
Fitted model for simulated data for distance travelled versus time of falling object measured with error.

But we were pretending to be Galileo and so we don't know the parameters in the model. The data does suggest it is a parabola, so we model it as such:

$$Y_i = \beta_0 + \beta_1 x_i + \beta_2 x_i^2 + \varepsilon_i, i = 1, \ldots, n$$

How do we find the LSE?

The lm function In R we can fit this model by simply using the lm function. We will describe this function in detail later, but here is a preview:

```
tt2 <-tt^2
fit <- lm(y~tt+tt2)
summary(fit)$coef
```

```
##                 Estimate Std. Error      t value      Pr(>|t|)
## (Intercept) 57.1047803  0.4996845 114.281666 5.119823e-32
## tt          -0.4460393  0.6806757  -0.655289 5.190757e-01
## tt2         -4.7471698  0.1933701 -24.549662 1.767229e-17
```

It gives us the LSE, as well as standard errors and p-values.
Part of what we do in this section is to explain the mathematics behind this function.

The least squares estimate (LSE) Let's write a function that computes the RSS for any vector β:

```
rss <- function(Beta0,Beta1,Beta2){
  r <- y - (Beta0+Beta1*tt+Beta2*tt^2)
  return(sum(r^2))
}
```

So for any three dimensional vector we get an RSS. Here is a plot of the RSS as a function of β_2 when we keep the other two fixed:

```
Beta2s<- seq(-10,0,len=100)
plot(Beta2s,sapply(Beta2s,rss,Beta0=55,Beta1=0),
     ylab="RSS",xlab="Beta2",type="l")
##Let's add another curve fixing another pair:
Beta2s<- seq(-10,0,len=100)
lines(Beta2s,sapply(Beta2s,rss,Beta0=65,Beta1=0),col=2)
```

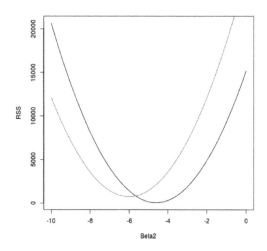

FIGURE 4.5
Residual sum of squares obtained for several values of the parameters.

Trial and error here is not going to work. Instead, we can use calculus: take the partial derivatives, set them to 0 and solve. Of course, if we have many parameters, these equations can get rather complex. Linear algebra provides a compact and general way of solving this problem.

More on Galton (Advanced) When studying the father-son data, Galton made a fascinating discovery using exploratory analysis.

He noted that if he tabulated the number of father-son height pairs and followed all the x,y values having the same totals in the table, they formed an ellipse. In the plot above, made by Galton, you see the ellipse formed by the pairs having 3 cases. This then led to modeling this data as correlated bivariate normal which we described earlier:

$$Pr(X < a, Y < b) = \int_{-\infty}^{a} \int_{-\infty}^{b} \frac{1}{2\pi\sigma_x\sigma_y\sqrt{1-\rho^2}}$$

$$\exp\left\{\frac{1}{2(1-\rho^2)}\left[\left(\frac{x-\mu_x}{\sigma_x}\right)^2 - 2\rho\left(\frac{x-\mu_x}{\sigma_x}\right)\left(\frac{y-\mu_y}{\sigma_y}\right) + \left(\frac{y-\mu_y}{\sigma_y}\right)^2\right]\right\}$$

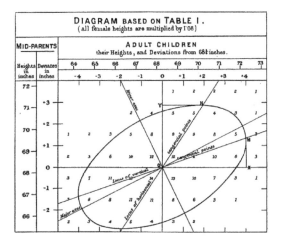

FIGURE 4.6
Galton's plot.

We described how we can use math to show that if you keep X fixed (condition to be x) the distribution of Y is normally distributed with mean: $\mu_x + \sigma_y \rho \left(\frac{x - \mu_x}{\sigma_x} \right)$ and standard deviation $\sigma_y \sqrt{1 - \rho^2}$. Note that ρ is the correlation between Y and X, which implies that if we fix $X = x$, Y does in fact follow a linear model. The β_0 and β_1 parameters in our simple linear model can be expressed in terms of $\mu_x, \mu_y, \sigma_x, \sigma_y$, and ρ.

4.2   Exercises

Here we include some refresher questions. If you haven't done so already, install the library UsingR.

```
install.packages("UsingR")
```

Once you load it, you have access to Galton's father and son heights:

```
data("father.son",package="UsingR")
```

1. What is the average height of the sons (don't round off)?

2. One of the defining features of regression is that we stratify one variable based on others. In Statistics, we use the verb "condition". For example, the linear model for son and father heights answers the question: how tall do I expect a son to be if I condition on his father being x inches? The regression line answers this question for any x.

Using the `father.son` dataset described above, we want to know the expected height of sons, if we condition on the father being 71 inches. Create a list of son heights for sons that have fathers with heights of 71 inches, rounding to the nearest inch.

What is the mean of the son heights for fathers that have a height of 71 inches (don't round off your answer)? Hint: use the function **round** on the fathers' heights.

3. We say a statistical model is a linear model when we can write it as a linear combination of parameters and known covariates, plus random error terms. In the choices below, Y represents our observations, time t is our only covariate, unknown parameters are represented with letters a, b, c, d and measurement error is represented by ε. If t is known, then any transformation of t is also known. So, for example, both $Y = a + bt + \varepsilon$ and $Y = a + bf(t) + \varepsilon$ are linear models. Which of the following **cannot** be written as a linear model?

    (a) $Y = a + bt + \varepsilon$

    (b) $Y = a + b\cos(t) + \varepsilon$

    (c) $Y = a + b^t + \varepsilon$

    (d) $Y = a + bt + ct^2 + dt^3 + \varepsilon$

4. Suppose you model the relationship between weight and height across individuals with a linear model. You assume that the height of individuals for a fixed weight x follows a linear model $Y = a + bx + \varepsilon$. Which of the following do you feel best describes what ε represents?

    (a) Measurement error: scales are not perfect.

    (b) Within individual random fluctuations: you don't weigh the same in the morning as in the afternoon.

    (c) Round off error introduced by the computer we use to analyze the data.

    (d) Between individual variability: people of the same height vary in their weight.

4.3   Matrix Notation

Here we introduce the basics of matrix notation. Initially this may seem over-complicated, but once we discuss examples, you will appreciate the power of using this notation to both explain and derive solutions, as well as implement them as R code.

The language of linear models Linear algebra notation actually simplifies the mathematical descriptions and manipulations of linear models, as well as coding in R. We will discuss the basics of this notation and then show some examples in R.

The main point of this entire exercise is to show how we can write the models above using matrix notation, and then explain how this is useful for solving the least squares equation. We start by simply defining notation and matrix multiplication, but bear with us since we eventually get back to the practical application.

4.4   Solving Systems of Equations

Linear algebra was created by mathematicians to solve systems of linear equations such as this:

$$a + b + c = 6$$
$$3a - 2b + c = 2$$
$$2a + b - c = 1$$

It provides very useful machinery to solve these problems generally. We will learn how we can write and solve this system using matrix algebra notation:

$$\begin{pmatrix} 1 & 1 & 1 \\ 3 & -2 & 1 \\ 2 & 1 & -1 \end{pmatrix} \begin{pmatrix} a \\ b \\ c \end{pmatrix} = \begin{pmatrix} 6 \\ 2 \\ 1 \end{pmatrix} \implies \begin{pmatrix} a \\ b \\ c \end{pmatrix} = \begin{pmatrix} 1 & 1 & 1 \\ 3 & -2 & 1 \\ 2 & 1 & -1 \end{pmatrix}^{-1} \begin{pmatrix} 6 \\ 2 \\ 1 \end{pmatrix}$$

This section explains the notation used above. It turns out that we can borrow this notation for linear models in statistics as well.

4.5   Vectors, Matrices, and Scalars

In the falling object, father-son heights, and mouse weight examples, the random variables associated with the data were represented by $Y_1, \ldots, Y_n$. We can think of this as a vector. In fact, in R we are already doing this:

```
data(father.son,package="UsingR")
y=father.son$fheight
head(y)
```

```
## [1] 65.04851 63.25094 64.95532 65.75250 61.13723 63.02254
```

In math we can also use just one symbol. We usually use bold to distinguish it from the individual entries:

$$\mathbf{Y} = \begin{pmatrix} Y_1 \\ Y_2 \\ \vdots \\ Y_N \end{pmatrix}$$

For reasons that will soon become clear, default representation of data vectors have dimension $N \times 1$ as opposed to $1 \times N$.

Here we don't always use bold because normally one can tell what is a matrix from the context.

Similarly, we can use math notation to represent the covariates or predictors. In a case with two predictors we can represent them like this:

$$\mathbf{X}_1 = \begin{pmatrix} x_{1,1} \\ \vdots \\ x_{N,1} \end{pmatrix} \text{ and } \mathbf{X}_2 = \begin{pmatrix} x_{1,2} \\ \vdots \\ x_{N,2} \end{pmatrix}$$

Note that for the falling object example $x_{1,1} = t_i$ and $x_{1,1} = t_i^2$ with t_i the time of the i-th observation. Also, keep in mind that vectors can be thought of as $N \times 1$ matrices.

For reasons that will soon become apparent, it is convenient to represent these in matrices:

$$\mathbf{X} = [\mathbf{X}_1 \mathbf{X}_2] = \begin{pmatrix} x_{1,1} & x_{1,2} \\ \vdots & \\ x_{N,1} & x_{N,2} \end{pmatrix}$$

This matrix has dimension $N \times 2$. We can create this matrix in R this way:

```
n <- 25
tt <- seq(0,3.4,len=n) ##time in secs, t is a base function
X <- cbind(X1=tt,X2=tt^2)
head(X)
```

```
##               X1         X2
## [1,] 0.0000000 0.00000000
## [2,] 0.1416667 0.02006944
## [3,] 0.2833333 0.08027778
## [4,] 0.4250000 0.18062500
## [5,] 0.5666667 0.32111111
## [6,] 0.7083333 0.50173611
```

```
dim(X)
```

```
## [1] 25   2
```

We can also use this notation to denote an arbitrary number of covariates with the following $N \times p$ matrix:

$$\mathbf{X} = \begin{pmatrix} x_{1,1} & \cdots & x_{1,p} \\ x_{2,1} & \cdots & x_{2,p} \\ & \vdots & \\ x_{N,1} & \cdots & x_{N,p} \end{pmatrix}$$

Just as an example, we show you how to make one in R now using `matrix` instead of `cbind`:

```
N <- 100; p <- 5
X <- matrix(1:(N*p),N,p)
head(X)
```

```
##      [,1] [,2] [,3] [,4] [,5]
## [1,]    1  101  201  301  401
## [2,]    2  102  202  302  402
## [3,]    3  103  203  303  403
## [4,]    4  104  204  304  404
## [5,]    5  105  205  305  405
## [6,]    6  106  206  306  406
```

```
dim(X)
```

```
## [1] 100   5
```

By default, the matrices are filled column by column. The `byrow=TRUE` argument lets us change that to row by row:

```
N <- 100; p <- 5
X <- matrix(1:(N*p),N,p,byrow=TRUE)
head(X)
```

```
##       [,1] [,2] [,3] [,4] [,5]
## [1,]    1    2    3    4    5
## [2,]    6    7    8    9   10
## [3,]   11   12   13   14   15
## [4,]   16   17   18   19   20
## [5,]   21   22   23   24   25
## [6,]   26   27   28   29   30
```

Finally, we define a scalar. A scalar is just a number, which we call a scalar because we want to distinguish it from vectors and matrices. We usually use lower case and don't bold. In the next section, we will understand why we make this distinction.

4.6 Exercises

1. In R we have vectors and matrices. You can create your own vectors with the function c.

```
c(1,5,3,4)
```

They are also the output of many functions such as:

```
rnorm(10)
```

You can turn vectors into matrices using functions such as rbind, cbind or matrix. Create the matrix from the vector 1:1000 like this:

```
X = matrix(1:1000,100,10)
```

What is the entry in row 25, column 3?

2. Using the function cbind, create a 10 x 5 matrix with first column x=1:10. Then add 2*x, 3*x, 4*x and 5*x to columns 2 through 5. What is the sum of the elements of the 7th row?

3. Which of the following creates a matrix with multiples of 3 in the third column?

- A) matrix(1:60,20,3)

- B) matrix(1:60,20,3,byrow=TRUE)

- C) x=11:20; rbind(x,2*x,3*x)

- D) x=1:40; matrix(3*x,20,2)

4.7 Matrix Operations

In a previous section, we motivated the use of matrix algebra with this system of equations:

$$a + b + c = 6$$
$$3a - 2b + c = 2$$
$$2a + b - c = 1$$

We described how this system can be rewritten and solved using matrix algebra:

$$\begin{pmatrix} 1 & 1 & 1 \\ 3 & -2 & 1 \\ 2 & 1 & -1 \end{pmatrix} \begin{pmatrix} a \\ b \\ c \end{pmatrix} = \begin{pmatrix} 6 \\ 2 \\ 1 \end{pmatrix} \implies \begin{pmatrix} a \\ b \\ c \end{pmatrix} = \begin{pmatrix} 1 & 1 & 1 \\ 3 & -2 & 1 \\ 2 & 1 & -1 \end{pmatrix}^{-1} \begin{pmatrix} 6 \\ 2 \\ 1 \end{pmatrix}$$

Having described matrix notation, we will explain the operation we perform with them. For example, above we have matrix multiplication and we also have a symbol representing the inverse of a matrix. The importance of these operations and others will become clear once we present specific examples related to data analysis.

Multiplying by a scalar We start with one of the simplest operations: scalar multiplication. If a is scalar and $\mathbf{X}$ is a matrix, then:

$$\mathbf{X} = \begin{pmatrix} x_{1,1} & \cdots & x_{1,p} \\ x_{2,1} & \cdots & x_{2,p} \\ & \vdots & \\ x_{N,1} & \cdots & x_{N,p} \end{pmatrix} \implies a\mathbf{X} = \begin{pmatrix} ax_{1,1} & \cdots & ax_{1,p} \\ ax_{2,1} & \cdots & ax_{2,p} \\ & \vdots & \\ ax_{N,1} & \cdots & ax_{N,p} \end{pmatrix}$$

R automatically follows this rule when we multiply a number by a matrix using *:

```
X <- matrix(1:12,4,3)
print(X)
```

```
##      [,1] [,2] [,3]
## [1,]    1    5    9
## [2,]    2    6   10
## [3,]    3    7   11
## [4,]    4    8   12
```

```
a <- 2
print(a*X)
```

```
##      [,1] [,2] [,3]
## [1,]    2   10   18
## [2,]    4   12   20
## [3,]    6   14   22
## [4,]    8   16   24
```

The transpose The transpose is an operation that simply changes columns to rows. We use a $\top$ to denote a transpose. The technical definition is as follows: if X is as we defined it above, here is the transpose which will be $p \times N$:

$$\mathbf{X} = \begin{pmatrix} x_{1,1} & \cdots & x_{1,p} \\ x_{2,1} & \cdots & x_{2,p} \\ & \vdots & \\ x_{N,1} & \cdots & x_{N,p} \end{pmatrix} \implies \mathbf{X}^{\top} = \begin{pmatrix} x_{1,1} & \cdots & x_{p,1} \\ x_{1,2} & \cdots & x_{p,2} \\ & \vdots & \\ x_{1,N} & \cdots & x_{p,N} \end{pmatrix}$$

In R we simply use t:

```
X <- matrix(1:12,4,3)
X
```

```
##      [,1] [,2] [,3]
## [1,]   1   5    9
## [2,]   2   6   10
## [3,]   3   7   11
## [4,]   4   8   12
```

```
t(X)
```

```
##      [,1] [,2] [,3] [,4]
## [1,]   1   2    3    4
## [2,]   5   6    7    8
## [3,]   9  10   11   12
```

Matrix multiplication We start by describing the matrix multiplication shown in the original system of equations example:

$$a + b + c = 6$$
$$3a - 2b + c = 2$$
$$2a + b - c = 1$$

What we are doing is multiplying the rows of the first matrix by the columns of the second. Since the second matrix only has one column, we perform this multiplication by doing the following:

$$\begin{pmatrix} 1 & 1 & 1 \\ 3 & -2 & 1 \\ 2 & 1 & -1 \end{pmatrix} \begin{pmatrix} a \\ b \\ c \end{pmatrix} = \begin{pmatrix} a+b+c \\ 3a-2b+c \\ 2a+b-c \end{pmatrix}$$

Here is a simple example. We can check to see if abc=c(3,2,1) is a solution:

```
X   <- matrix(c(1,3,2,1,-2,1,1,1,-1),3,3)
abc <- c(3,2,1) #use as an example
rbind( sum(X[1,]*abc), sum(X[2,]*abc), sum(X[3,]*abc))
```

```
##      [,1]
## [1,]   6
## [2,]   6
## [3,]   7
```

We can use the %*% to perform the matrix multiplication and make this much more compact:

```
X%*%abc
```

```
##      [,1]
## [1,]   6
## [2,]   6
## [3,]   7
```

We can see that c(3,2,1) is not a solution as the answer here is not the required c(6,2,1).

To get the solution, we will need to invert the matrix on the left, a concept we learn about below.

Here is the general definition of matrix multiplication of matrices A and X:

$$\mathbf{AX} = \begin{pmatrix} a_{1,1} & a_{1,2} & \cdots & a_{1,N} \\ a_{2,1} & a_{2,2} & \cdots & a_{2,N} \\ & & \vdots & \\ a_{M,1} & a_{M,2} & \cdots & a_{M,N} \end{pmatrix} \begin{pmatrix} x_{1,1} & \cdots & x_{1,p} \\ x_{2,1} & \cdots & x_{2,p} \\ & \vdots & \\ x_{N,1} & \cdots & x_{N,p} \end{pmatrix}$$

$$= \begin{pmatrix} \sum_{i=1}^{N} a_{1,i} x_{i,1} & \cdots & \sum_{i=1}^{N} a_{1,i} x_{i,p} \\ & \vdots & \\ \sum_{i=1}^{N} a_{M,i} x_{i,1} & \cdots & \sum_{i=1}^{N} a_{M,i} x_{i,p} \end{pmatrix}$$

You can only take the product if the number of columns of the first matrix A equals the number of rows of the second one X. Also, the final matrix has the same row numbers as the first A and the same column numbers as the second X. After you study the example below, you may want to come back and re-read the sections above.

The identity matrix The identity matrix is analogous to the number 1: if you multiply the identity matrix by another matrix, you get the same matrix. For this to happen, we need it to be like this:

$$\mathbf{I} = \begin{pmatrix} 1 & 0 & 0 & \cdots & 0 & 0 \\ 0 & 1 & 0 & \cdots & 0 & 0 \\ 0 & 0 & 1 & \cdots & 0 & 0 \\ \vdots & \vdots & \vdots & \ddots & \vdots & \vdots \\ 0 & 0 & 0 & \cdots & 1 & 0 \\ 0 & 0 & 0 & \cdots & 0 & 1 \end{pmatrix}$$

By this definition, the identity always has to have the same number of rows as columns or be what we call a square matrix.

If you follow the matrix multiplication rule above, you notice this works out:

$$\mathbf{XI} = \begin{pmatrix} x_{1,1} & \cdots & x_{1,p} \\ & \vdots & \\ x_{N,1} & \cdots & x_{N,p} \end{pmatrix} \begin{pmatrix} 1 & 0 & 0 & \cdots & 0 & 0 \\ 0 & 1 & 0 & \cdots & 0 & 0 \\ 0 & 0 & 1 & \cdots & 0 & 0 \\ & & & \vdots & & \\ 0 & 0 & 0 & \cdots & 1 & 0 \\ 0 & 0 & 0 & \cdots & 0 & 1 \end{pmatrix} = \begin{pmatrix} x_{1,1} & \cdots & x_{1,p} \\ & \vdots & \\ x_{N,1} & \cdots & x_{N,p} \end{pmatrix}$$

In R you can form an identity matrix this way:

```
n <- 5 #pick dimensions
diag(n)
```

```
##      [,1] [,2] [,3] [,4] [,5]
## [1,]    1    0    0    0    0
## [2,]    0    1    0    0    0
## [3,]    0    0    1    0    0
## [4,]    0    0    0    1    0
## [5,]    0    0    0    0    1
```

The inverse The inverse of matrix X, denoted with X^{-1}, has the property that, when multiplied, gives you the identity $X^{-1}X = I$. Of course, not all matrices have inverses. For example, a 2×2 matrix with 1s in all its entries does not have an inverse.

As we will see when we get to the section on applications to linear models, being able to compute the inverse of a matrix is quite useful. A very convenient aspect of R is that it includes a predefined function `solve` to do this. Here is how we would use it to solve the linear of equations:

```
X <- matrix(c(1,3,2,1,-2,1,1,1,-1),3,3)
y <- matrix(c(6,2,1),3,1)
solve(X)%*%y #equivalent to solve(X,y)
```

```
##      [,1]
## [1,]    1
## [2,]    2
## [3,]    3
```

Please note that `solve` is a function that should be used with caution as it is not generally numerically stable. We explain this in much more detail in the QR factorization section.

4.8 Exercises

1. Suppose X is a matrix in R. Which of the following is **not** equivalent to X?
   (a) `t( t(X) )`
   (b) `X %*% matrix(1,ncol(X) )`
   (c) `X*1`
   (d) `X%*%diag(ncol(X))`

2. Solve the following system of equations using R:

$$3a + 4b - 5c + d = 10$$
$$2a + 2b + 2c - d = 5$$
$$a - b + 5c - 5d = 7$$
$$5a + d = 4$$

What is the solution for c?

3. Load the following two matrices into R:

```
a <- matrix(1:12, nrow=4)
b <- matrix(1:15, nrow=3)
```

Note the dimension of a and the dimension of b.

In the question below, we will use the matrix multiplication operator in R, `%*%`, to multiply these two matrices.

What is the value in the 3rd row and the 2nd column of the matrix product of a and b?

4. Multiply the 3rd row of a with the 2nd column of b, using the element-wise vector multiplication with `*`.

What is the sum of the elements in the resulting vector?

4.9 Examples

Now we are ready to see how matrix algebra can be useful when analyzing data. We start with some simple examples and eventually arrive at the main one: how to write linear models with matrix algebra notation and solve the least squares problem.

The average To compute the sample average and variance of our data, we use these formulas $\bar{Y} = \frac{1}{N}Y_i$ and $\text{var}(Y) = \frac{1}{N}\sum_{i=1}^{N}(Y_i - \bar{Y})^2$. We can represent these with matrix multiplication. First, define this $N \times 1$ matrix made just of 1s:

$$A = \begin{pmatrix} 1 \\ 1 \\ \vdots \\ 1 \end{pmatrix}$$

This implies that:

$$\frac{1}{N}\mathbf{A}^\top Y = \frac{1}{N}\begin{pmatrix} 1 & 1 & ,\dots & 1 \end{pmatrix}\begin{pmatrix} Y_1 \\ Y_2 \\ \vdots \\ Y_N \end{pmatrix} = \frac{1}{N}\sum_{i=1}^{N}Y_i = \bar{Y}$$

Note that we are multiplying by the scalar $1/N$. In R, we multiply matrix using %*%:

```
data(father.son,package="UsingR")
y <- father.son$sheight
print(mean(y))
```

```
## [1] 68.68407
```

```
N <- length(y)
Y<- matrix(y,N,1)
A <- matrix(1,N,1)
barY=t(A)%*%Y / N
```

```
print(barY)
```

```
##              [,1]
## [1,] 68.68407
```

The variance As we will see later, multiplying the transpose of a matrix with another is very common in statistics. In fact, it is so common that there is a function in R:

```
barY=crossprod(A,Y) / N
print(barY)
```

```
##              [,1]
## [1,] 68.68407
```

For the variance, we note that if:

$$\mathbf{r} \equiv \begin{pmatrix} Y_1 - \bar{Y} \\ \vdots \\ Y_N - \bar{Y} \end{pmatrix}, \quad \frac{1}{N}\mathbf{r}^\top\mathbf{r} = \frac{1}{N}\sum_{i=1}^{N}(Y_i - \bar{Y})^2$$

In R, if you only send one matrix into `crossprod`, it computes: $r^\top r$ so we can simply type:

```
r <- y - barY
crossprod(r)/N
```

```
##           [,1]
## [1,] 7.915196
```

Which is almost equivalent to:

```
library(rafalib)
popvar(y)
```

```
## [1] 7.915196
```

Linear models Now we are ready to put all this to use. Let's start with Galton's example. If we define these matrices:

$$\mathbf{Y} = \begin{pmatrix} Y_1 \\ Y_2 \\ \vdots \\ Y_N \end{pmatrix}, \mathbf{X} = \begin{pmatrix} 1 & x_1 \\ 1 & x_2 \\ \vdots & \\ 1 & x_N \end{pmatrix}, \beta = \begin{pmatrix} \beta_0 \\ \beta_1 \end{pmatrix} \text{ and } \varepsilon = \begin{pmatrix} \varepsilon_1 \\ \varepsilon_2 \\ \vdots \\ \varepsilon_N \end{pmatrix}$$

Then we can write the model:

$$Y_i = \beta_0 + \beta_1 x_i + \varepsilon_i, i = 1, \ldots, N$$

as:

$$\begin{pmatrix} Y_1 \\ Y_2 \\ \vdots \\ Y_N \end{pmatrix} = \begin{pmatrix} 1 & x_1 \\ 1 & x_2 \\ \vdots & \\ 1 & x_N \end{pmatrix} \begin{pmatrix} \beta_0 \\ \beta_1 \end{pmatrix} + \begin{pmatrix} \varepsilon_1 \\ \varepsilon_2 \\ \vdots \\ \varepsilon_N \end{pmatrix}$$

or simply:

$$\mathbf{Y} = \mathbf{X}\beta + \varepsilon$$

which is a much simpler way to write it.

The least squares equation becomes simpler as well since it is the following cross-product:

$$(\mathbf{Y} - \mathbf{X}\beta)^\top (\mathbf{Y} - \mathbf{X}\beta)$$

So now we are ready to determine which values of β minimize the above, which we can do using calculus to find the minimum.

Advanced: Finding the minimum using calculus There are a series of rules that permit us to compute partial derivative equations in matrix notation. By equating the derivative to 0 and solving for the β, we will have our solution. The only one we need here tells us that the derivative of the above equation is:

$$2\mathbf{X}^\top(\mathbf{Y} - \mathbf{X}\hat{\beta}) = 0$$

$$\mathbf{X}^\top\mathbf{X}\hat{\beta} = \mathbf{X}^\top\mathbf{Y}$$

$$\hat{\beta} = (\mathbf{X}^\top\mathbf{X})^{-1}\mathbf{X}^\top\mathbf{Y}$$

and we have our solution. We usually put a hat on the β that solves this, $\hat{\beta}$, as it is an estimate of the "real" β that generated the data.

Remember that the least squares are like a square (multiply something by itself) and that this formula is similar to the derivative of $f(x)^2$ being $2f(x)f'(x)$.

Finding LSE in R Let's see how it works in R:

```
data(father.son,package="UsingR")
x=father.son$fheight
y=father.son$sheight
X <- cbind(1,x)
betahat <- solve( t(X) %*% X ) %*% t(X) %*% y
###or
betahat <- solve( crossprod(X) ) %*% crossprod( X, y )
```

Now we can see the results of this by computing the estimated $\hat{\beta}_0 + \hat{\beta}_1 x$ for any value of x:

```
newx <- seq(min(x),max(x),len=100)
X <- cbind(1,newx)
fitted <- X%*%betahat
plot(x,y,xlab="Father's height",ylab="Son's height")
lines(newx,fitted,col=2)
```

This $\hat{\beta} = (\mathbf{X}^\top\mathbf{X})^{-1}\mathbf{X}^\top\mathbf{Y}$ is one of the most widely used results in data analysis. One of the advantages of this approach is that we can use it in many different situations. For example, in our falling object problem:

```
set.seed(1)
g <- 9.8 #meters per second
n <- 25
tt <- seq(0,3.4,len=n) #time in secs, t is a base function
d <- 56.67  - 0.5*g*tt^2 + rnorm(n,sd=1)
```

Notice that we are using almost the same exact code:

```
X <- cbind(1,tt,tt^2)
y <- d
betahat <- solve(crossprod(X))%*%crossprod(X,y)
newtt <- seq(min(tt),max(tt),len=100)
X <- cbind(1,newtt,newtt^2)
```

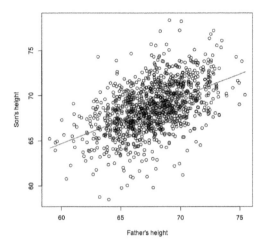

FIGURE 4.7
Galton's data with fitted regression line.

```
fitted <- X%*%betahat
plot(tt,y,xlab="Time",ylab="Height")
lines(newtt,fitted,col=2)
```

And the resulting estimates are what we expect:

```
betahat
```

```
##           [,1]
##     56.5317368
## tt   0.5013565
##     -5.0386455
```

The Tower of Pisa is about 56 meters high. Since we are just dropping the object there is no initial velocity, and half the constant of gravity is 9.8/2=4.9 meters per second squared.

The `lm` Function R has a very convenient function that fits these models. We will learn more about this function later, but here is a preview:

```
X <- cbind(tt,tt^2)
fit=lm(y~X)
summary(fit)
```

```
##
## Call:
## lm(formula = y ~ X)
##
## Residuals:
##     Min      1Q  Median      3Q     Max
## -2.5295 -0.4882  0.2537  0.6560  1.5455
##
```

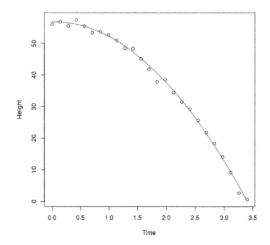

FIGURE 4.8
Fitted parabola to simulated data for distance travelled versus time of falling object measured with error.

```
## Coefficients:
##                Estimate Std. Error t value Pr(>|t|)
## (Intercept)    56.5317     0.5451 103.701    <2e-16 ***
## Xtt             0.5014     0.7426   0.675     0.507
## X              -5.0386     0.2110 -23.884    <2e-16 ***
## ---
## Signif. codes:  0 '***' 0.001 '**' 0.01 '*' 0.05 '.' 0.1 ' ' 1
##
## Residual standard error: 0.9822 on 22 degrees of freedom
## Multiple R-squared:  0.9973, Adjusted R-squared:  0.997
## F-statistic:  4025 on 2 and 22 DF,  p-value: < 2.2e-16
```

Note that we obtain the same values as above.

Summary We have shown how to write linear models using linear algebra. We are going to do this for several examples, many of which are related to designed experiments. We also demonstrated how to obtain least squares estimates. Nevertheless, it is important to remember that because y is a random variable, these estimates are random as well. In a later section, we will learn how to compute standard error for these estimates and use this to perform inference. {pagebreak}

4.10   Exercises

1. Suppose we are analyzing a set of 4 samples. The first two samples are from a treatment group A and the second two samples are from a treatment group B. This design can be represented with a model matrix like so:

```
X <- matrix(c(1,1,1,1,0,0,1,1),nrow=4)
rownames(X) <- c("a","a","b","b")
X
```

```
##    [,1] [,2]
## a    1    0
## a    1    0
## b    1    1
## b    1    1
```

Suppose that the fitted parameters for a linear model give us:

```
beta <- c(5, 2)
```

Use the matrix multiplication operator, %*%, in R to answer the following questions: What is the fitted value for the A samples? (The fitted Y values.)

2. What is the fitted value for the B samples? (The fitted Y values.)

3. Suppose now we are comparing two treatments B and C to a control group A, each with two samples. This design can be represented with a model matrix like so:

```
X <- matrix(c(1,1,1,1,1,1,0,0,1,1,0,0,0,0,0,0,1,1),nrow=6)
rownames(X) <- c("a","a","b","b","c","c")
X
```

Suppose that the fitted values for the linear model are given by:

```
beta <- c(10,3,-3)
```

What is the fitted value for the B samples?

4. What is the fitted value for the C samples?

5

Linear Models

Many of the models we use in data analysis can be presented using matrix algebra. We refer to these types of models as *linear models*. "Linear" here does not refer to lines, but rather to linear combinations. The representations we describe are convenient because we can write models more succinctly and we have the matrix algebra mathematical machinery to facilitate computation. In this chapter, we will describe in some detail how we use matrix algebra to represent and fit.

In this book, we focus on linear models that represent dichotomous groups: treatment versus control, for example. The effect of diet on mice weights is an example of this type of linear model. Here we describe slightly more complicated models, but continue to focus on dichotomous variables.

As we learn about linear models, we need to remember that we are still working with random variables. This means that the estimates we obtain using linear models are also random variables. Although the mathematics is more complex, the concepts we learned in previous chapters apply here. We begin with some exercises to review the concept of random variables in the context of linear models.

5.1 Exercises

The standard error of an estimate is the standard deviation of the sampling distribution of an estimate. In previous chapters, we saw that our estimate of the mean of a population changed depending on the sample that we took from the population. If we repeatedly sampled from the population and each time estimated the mean, the collection of mean estimates would form the sampling distribution of the estimate. When we took the standard deviation of those estimates, that was the standard error of our mean estimate.

In the case of a linear model written as:

$$Y_i = \beta_0 + \beta_1 X_i + \varepsilon_i, i = 1, \ldots, n$$

ε_i is considered random. Every time we re-run the experiment, we will see different ε_i. This implies that in different application ε_i represents different things: measurement error or variability between individuals for example.

If we were to re-run the experiment many times and estimate linear model terms $\hat{\beta}$ each time, the distribution of these $\hat{\beta}$ is called the sampling distribution of the estimates. If we take the standard deviation of all of these estimates from repetitions of the experiment, this is called the standard error of the estimate. While we are not necessarily sampling individuals, you can think about the repetition of the experiment as "sampling" new errors in our observation of Y.

1. We have shown how to find the least squares estimates with matrix algebra. These estimates are random variables as they are linear combinations of the data. For

these estimates to be useful, we also need to compute the standard errors. Here we review standard errors in the context of linear models. To see this, we can run a Monte Carlo simulation to imitate the collection of falling object data. Specifically, we will generate the data repeatedly and compute the estimate for the quadratic term each time.

```
g = 9.8
h0 = 56.67
v0 = 0
n = 25
tt = seq(0,3.4,len=n)
y = h0 + v0 *tt - 0.5* g*tt^2 + rnorm(n,sd=1)
```

Now we act as if we didn't know h0, v0 and -0.5*g and use regression to estimate these. We can rewrite the model as $y = \beta_0 + \beta_1 t + \beta_2 t^2 + \varepsilon$ and obtain the LSE we have used in this class. Note that g = -2 β_2.

To obtain the LSE in R we could write:

```
X = cbind(1,tt,tt^2)
A = solve(crossprod(X))%*%t(X)
```

Given how we have defined A, which of the following is the LSE of g, the acceleration due to gravity? Hint: try the code in R.

- A) 9.8

- B) A %*% y

- C) -2 * (A %*% y) [3]

- D) A[3,3]

2. In the lines of code above, the function **rnorm** introduced randomness. This means that each time the lines of code above are repeated, the estimate of g will be different.

Use the code above in conjunction with the function **replicate** to generate 100,000 Monte Carlo simulated datasets. For each dataset, compute an estimate of g. (Remember to multiply by -2.)

What is the standard error of this estimate?

3. In the father and son height examples, we have randomness because we have a random sample of father and son pairs. For the sake of illustration, let's assume that this is the entire population:

```
library(UsingR)
x = father.son$fheight
y = father.son$sheight
n = length(y)
```

Now let's run a Monte Carlo simulation in which we take a sample of size 50 over and over again. Here is how we obtain one sample:

```
N = 50
index = sample(n,N)
sampledat = father.son[index,]
x = sampledat$fheight
y = sampledat$sheight
betahat = lm(y~x)$coef
```

Use the function `replicate` to take 10,000 samples.

What is the standard error of the slope estimate? That is, calculate the standard deviation of the estimate from the observed values obtained from many random samples.

4. Later in this chapter we will introduce a new concept: covariance. The covariance of two lists of numbers $X = x_1, ..., x_n$ and $Y = y_1, ..., y_n$ is:

```
n <- 100
Y <- rnorm(n)
X <- rnorm(n)
mean( (Y - mean(Y))*(X-mean(X) ) )
```

Which of the following is closest to the covariance between father heights and son heights?

- A) 0

- B) -4

- C) 4

- D) 0.5

5.2   The Design Matrix

Here we will show how to use the two R functions, `formula` and `model.matrix`, in order to produce *design matrices* (also known as *model matrices*) for a variety of linear models. For example, in the mouse diet examples we wrote the model as

$$Y_i = \beta_0 + \beta_1 x_i + \varepsilon_i, i = 1, \ldots, N$$

with Y_i the weights and x_i equal to 1 only when mouse i receives the high fat diet. We use the term *experimental unit* to N different entities from which we obtain a measurement. In this case, the mice are the experimental units.

This is the type of variable we will focus on in this chapter. We call them *indicator variables* since they simply indicate if the experimental unit had a certain characteristic or not. As we described earlier, we can use linear algebra to represent this model:

$$\mathbf{Y} = \begin{pmatrix} Y_1 \\ Y_2 \\ \vdots \\ Y_N \end{pmatrix}, \mathbf{X} = \begin{pmatrix} 1 & x_1 \\ 1 & x_2 \\ \vdots \\ 1 & x_N \end{pmatrix}, \boldsymbol{\beta} = \begin{pmatrix} \beta_0 \\ \beta_1 \end{pmatrix} \text{ and } \varepsilon = \begin{pmatrix} \varepsilon_1 \\ \varepsilon_2 \\ \vdots \\ \varepsilon_N \end{pmatrix}$$

as:

$$\begin{pmatrix} Y_1 \\ Y_2 \\ \vdots \\ Y_N \end{pmatrix} = \begin{pmatrix} 1 & x_1 \\ 1 & x_2 \\ \vdots & \\ 1 & x_N \end{pmatrix} \begin{pmatrix} \beta_0 \\ \beta_1 \end{pmatrix} + \begin{pmatrix} \varepsilon_1 \\ \varepsilon_2 \\ \vdots \\ \varepsilon_N \end{pmatrix}$$

or simply:

$$\mathbf{Y} = \mathbf{X}\boldsymbol{\beta} + \boldsymbol{\varepsilon}$$

The design matrix is the matrix $\mathbf{X}$.

Once we define a design matrix, we are ready to find the least squares estimates. We refer to this as *fitting the model*. For fitting linear models in R, we will directly provide a *formula* to the `lm` function. In this script, we will use the `model.matrix` function, which is used internally by the `lm` function. This will help us to connect the R `formula` with the matrix $\mathbf{X}$. It will therefore help us interpret the results from `lm`.

Choice of design The choice of design matrix is a critical step in linear modeling since it encodes which coefficients will be fit in the model, as well as the inter-relationship between the samples. A common misunderstanding is that the choice of design follows straightforward from a description of which samples were included in the experiment. This is not the case. The basic information about each sample (whether control or treatment group, experimental batch, etc.) does not imply a single 'correct' design matrix. The design matrix additionally encodes various assumptions about how the variables in $\mathbf{X}$ explain the observed values in $\mathbf{Y}$, on which the investigator must decide.

For the examples we cover here, we use linear models to make comparisons between different groups. Hence, the design matrices that we ultimately work with will have at least two columns: an *intercept* column, which consists of a column of 1's, and a second column, which specifies which samples are in a second group. In this case, two coefficients are fit in the linear model: the intercept, which represents the population average of the first group, and a second coefficient, which represents the difference between the population averages of the second group and the first group. The latter is typically the coefficient we are interested in when we are performing statistical tests: we want to know if there is a difference between the two groups.

We encode this experimental design in R with two pieces. We start with a formula with the tilde symbol ˜. This means that we want to model the observations using the variables to the right of the tilde. Then we put the name of a variable, which tells us which samples are in which group.

Let's try an example. Suppose we have two groups, control and high fat diet, with two samples each. For illustrative purposes, we will code these with 1 and 2 respectively. We should first tell R that these values should not be interpreted numerically, but as different levels of a *factor*. We can then use the paradigm ˜ group to, say, model on the variable group.

```
group <- factor( c(1,1,2,2) )
model.matrix(~ group)
```

```
##   (Intercept) group2
## 1           1      0
## 2           1      0
## 3           1      1
## 4           1      1
```

```
## attr(,"assign")
## [1] 0 1
## attr(,"contrasts")
## attr(,"contrasts")$group
## [1] "contr.treatment"
```

(Don't worry about the `attr` lines printed beneath the matrix. We won't be using this information.)

What about the `formula` function? We don't have to include this. By starting an expression with ~, it is equivalent to telling R that the expression is a formula:

```
model.matrix(formula(~ group))
```

```
##   (Intercept) group2
## 1           1      0
## 2           1      0
## 3           1      1
## 4           1      1
## attr(,"assign")
## [1] 0 1
## attr(,"contrasts")
## attr(,"contrasts")$group
## [1] "contr.treatment"
```

What happens if we don't tell R that `group` should be interpreted as a factor?

```
group <- c(1,1,2,2)
model.matrix(~ group)
```

```
##   (Intercept) group
## 1           1     1
## 2           1     1
## 3           1     2
## 4           1     2
## attr(,"assign")
## [1] 0 1
```

This is **not** the design matrix we wanted, and the reason is that we provided a numeric variable as opposed to an *indicator* to the `formula` and `model.matrix` functions, without saying that these numbers actually referred to different groups. We want the second column to have only 0 and 1, indicating group membership.

A note about factors: the names of the levels are irrelevant to `model.matrix` and `lm`. All that matters is the order. For example:

```
group <- factor(c("control","control","highfat","highfat"))
model.matrix(~ group)
```

```
##   (Intercept) grouphighfat
## 1           1            0
## 2           1            0
## 3           1            1
## 4           1            1
## attr(,"assign")
```

```
## [1] 0 1
## attr(,"contrasts")
## attr(,"contrasts")$group
## [1] "contr.treatment"
```

produces the same design matrix as our first code chunk.

More groups Using the same formula, we can accommodate modeling more groups. Suppose we have a third diet:

```
group <- factor(c(1,1,2,2,3,3))
model.matrix(~ group)
```

```
##   (Intercept) group2 group3
## 1           1      0      0
## 2           1      0      0
## 3           1      1      0
## 4           1      1      0
## 5           1      0      1
## 6           1      0      1
## attr(,"assign")
## [1] 0 1 1
## attr(,"contrasts")
## attr(,"contrasts")$group
## [1] "contr.treatment"
```

Now we have a third column which specifies which samples belong to the third group. An alternate formulation of design matrix is possible by specifying + 0 in the formula:

```
group <- factor(c(1,1,2,2,3,3))
model.matrix(~ group + 0)
```

```
##   group1 group2 group3
## 1      1      0      0
## 2      1      0      0
## 3      0      1      0
## 4      0      1      0
## 5      0      0      1
## 6      0      0      1
## attr(,"assign")
## [1] 1 1 1
## attr(,"contrasts")
## attr(,"contrasts")$group
## [1] "contr.treatment"
```

This group now fits a separate coefficient for each group. We will explore this design in more depth later on.

More variables We have been using a simple case with just one variable (diet) as an example. In the life sciences, it is quite common to perform experiments with more than one variable. For example, we may be interested in the effect of diet and the difference in sexes. In this case, we have four possible groups:

```
diet <- factor(c(1,1,1,1,2,2,2,2))
sex <- factor(c("f","f","m","m","f","f","m","m"))
table(diet,sex)
```

```
##      sex
## diet f m
##    1 2 2
##    2 2 2
```

If we assume that the diet effect is the same for males and females (this is an assumption), then our linear model is:

$$Y_i = \beta_0 + \beta_1 x_{i,1} + \beta_2 x_{i,2} + \varepsilon_i$$

To fit this model in R, we can simply add the additional variable with a + sign in order to build a design matrix which fits based on the information in additional variables:

```
diet <- factor(c(1,1,1,1,2,2,2,2))
sex <- factor(c("f","f","m","m","f","f","m","m"))
model.matrix(~ diet + sex)
```

```
##   (Intercept) diet2 sexm
## 1           1     0    0
## 2           1     0    0
## 3           1     0    1
## 4           1     0    1
## 5           1     1    0
## 6           1     1    0
## 7           1     1    1
## 8           1     1    1
## attr(,"assign")
## [1] 0 1 2
## attr(,"contrasts")
## attr(,"contrasts")$diet
## [1] "contr.treatment"
##
## attr(,"contrasts")$sex
## [1] "contr.treatment"
```

The design matrix includes an intercept, a term for `diet` and a term for `sex`. We would say that this linear model accounts for differences in both the group and condition variables. However, as mentioned above, the model assumes that the diet effect is the same for both males and females. We say these are an *additive* effect. For each variable, we add an effect regardless of what the other is. Another model is possible here, which fits an additional term and which encodes the potential interaction of group and condition variables. We will cover interaction terms in depth in a later script.

The interaction model can be written in either of the following two formulas:

```
model.matrix(~ diet + sex + diet:sex)
```

or

```
model.matrix(~ diet*sex)
```

```
##    (Intercept) diet2 sexm diet2:sexm
## 1            1     0    0           0
## 2            1     0    0           0
## 3            1     0    1           0
## 4            1     0    1           0
## 5            1     1    0           0
## 6            1     1    0           0
## 7            1     1    1           1
## 8            1     1    1           1
## attr(,"assign")
## [1] 0 1 2 3
## attr(,"contrasts")
## attr(,"contrasts")$diet
## [1] "contr.treatment"
##
## attr(,"contrasts")$sex
## [1] "contr.treatment"
```

Releveling   The level which is chosen for the *reference level* is the level which is contrasted against. By default, this is simply the first level alphabetically. We can specify that we want group 2 to be the reference level by either using the `relevel` function:

```
group <- factor(c(1,1,2,2))
group <- relevel(group, "2")
model.matrix(~ group)
```

```
##    (Intercept) group1
## 1            1      1
## 2            1      1
## 3            1      0
## 4            1      0
## attr(,"assign")
## [1] 0 1
## attr(,"contrasts")
## attr(,"contrasts")$group
## [1] "contr.treatment"
```

or by providing the levels explicitly in the `factor` call:

```
group <- factor(group, levels=c("1","2"))
model.matrix(~ group)
```

```
##    (Intercept) group2
## 1            1      0
## 2            1      0
## 3            1      1
## 4            1      1
## attr(,"assign")
## [1] 0 1
## attr(,"contrasts")
## attr(,"contrasts")$group
## [1] "contr.treatment"
```

Where does model.matrix look for the data? The `model.matrix` function will grab the variable from the R global environment, unless the data is explicitly provided as a data frame to the `data` argument:

```
group <- 1:4
model.matrix(~ group, data=data.frame(group=5:8))

##   (Intercept) group
## 1           1     5
## 2           1     6
## 3           1     7
## 4           1     8
## attr(,"assign")
## [1] 0 1
```

Note how the R global environment variable `group` is ignored.

Continuous variables In this chapter, we focus on models based on indicator values. In certain designs, however, we will be interested in using numeric variables in the design formula, as opposed to converting them to factors first. For example, in the falling object example, time was a continuous variable in the model and time squared was also included:

```
tt <- seq(0,3.4,len=4)
model.matrix(~ tt + I(tt^2))

##   (Intercept)       tt    I(tt^2)
## 1           1 0.000000  0.000000
## 2           1 1.133333  1.284444
## 3           1 2.266667  5.137778
## 4           1 3.400000 11.560000
## attr(,"assign")
## [1] 0 1 2
```

The `I` function above is necessary to specify a mathematical transformation of a variable. For more details, see the manual page for the `I` function by typing `?I`.

In the life sciences, we could be interested in testing various dosages of a treatment, where we expect a specific relationship between a measured quantity and the dosage, e.g. 0 mg, 10 mg, 20 mg.

The assumptions imposed by including continuous data as variables are typically hard to defend and motivate than the indicator function variables. Whereas the indicator variables simply assume a different mean between two groups, continuous variables assume a very specific relationship between the outcome and predictor variables.

In cases like the falling object, we have the theory of gravitation supporting the model. In the father-son height example, because the data is bivariate normal, it follows that there is a linear relationship if we condition. However, we find that continuous variables are included in linear models without justification to "adjust" for variables such as age. We highly discourage this practice unless the data support the model being used.

5.3   Exercises

Suppose we have an experiment with the following design: on three different days, we perform an experiment with two treated and two control units. We then measure some outcome Y_i, and we want to test the effect of treatment as well as the effects of different days (perhaps the temperature in the lab affects the measuring device). Assume that the true condition effect is the same for each day (no interaction between condition and day). We then define factors in R for `day` and for `condition`.

condition/day	A	B	C
treatment	2	2	2
control	2	2	2

1. Given the factors we have defined above and without defining any new ones, which of the following R formula will produce a design matrix (model matrix) that lets us analyze the effect of condition, controlling for the different days?

- A) ~ day + condition

- B) ~ condition   ~ day

- C) ~ A + B + C + control + treated

- D) ~ B + C + treated

Remember that using the ~ and the names for the two variables we want in the model will produce a design matrix controlling for all levels of day and all levels of condition. We do not use the levels in the design formula.

The mouse diet example We will demonstrate how to analyze the high fat diet data using linear models instead of directly applying a t-test. We will demonstrate how ultimately these two approaches are equivalent.

We start by reading in the data and creating a quick stripchart:

```
dat <- read.csv("femaleMiceWeights.csv") ##previously downloaded
stripchart(dat$Bodyweight ~ dat$Diet, vertical=TRUE, method="jitter",
           main="Bodyweight over Diet")
```

We can see that the high fat diet group appears to have higher weights on average, although there is overlap between the two samples.

For demonstration purposes, we will build the design matrix **X** using the formula ~ Diet. The group with the 1's in the second column is determined by the level of Diet which comes second; that is, the non-reference level.

```
levels(dat$Diet)
```

```
## [1] "chow" "hf"
```

```
X <- model.matrix(~ Diet, data=dat)
head(X)
```

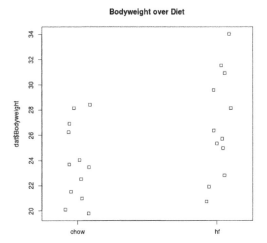

FIGURE 5.1
Mice bodyweights stratified by diet.

```
##   (Intercept) Diethf
## 1           1      0
## 2           1      0
## 3           1      0
## 4           1      0
## 5           1      0
## 6           1      0
```

5.4   The Mathematics Behind lm()

Before we use our shortcut for running linear models, lm, we want to review what will happen internally. Inside of lm, we will form the design matrix $\mathbf{X}$ and calculate the β, which minimizes the sum of squares using the previously described formula. The formula for this solution is:

$$\hat{\beta} = (\mathbf{X}^\top \mathbf{X})^{-1} \mathbf{X}^\top \mathbf{Y}$$

We can calculate this in R using our matrix multiplication operator %*%, the inverse function `solve`, and the transpose function `t`.

```
Y <- dat$Bodyweight
X <- model.matrix(~ Diet, data=dat)
solve(t(X) %*% X) %*% t(X) %*% Y
```

```
##                   [,1]
## (Intercept) 23.813333
## Diethf       3.020833
```

These coefficients are the average of the control group and the difference of the averages:

```
s <- split(dat$Bodyweight, dat$Diet)
mean(s[["chow"]])
```

```
## [1] 23.81333
```

```
mean(s[["hf"]]) - mean(s[["chow"]])
```

```
## [1] 3.020833
```

Finally, we use our shortcut, lm, to run the linear model:

```
fit <- lm(Bodyweight ~ Diet, data=dat)
summary(fit)
```

```
##
## Call:
## lm(formula = Bodyweight ~ Diet, data = dat)
##
## Residuals:
##     Min      1Q  Median      3Q     Max
## -6.1042 -2.4358 -0.4138  2.8335  7.1858
##
## Coefficients:
##             Estimate Std. Error t value Pr(>|t|)
## (Intercept)   23.813      1.039  22.912   <2e-16 ***
## Diethf         3.021      1.470   2.055   0.0519 .
## ---
## Signif. codes:  0 '***' 0.001 '**' 0.01 '*' 0.05 '.' 0.1 ' ' 1
##
## Residual standard error: 3.6 on 22 degrees of freedom
## Multiple R-squared:  0.1611, Adjusted R-squared:  0.1229
## F-statistic: 4.224 on 1 and 22 DF,  p-value: 0.05192
```

```
(coefs <- coef(fit))
```

```
## (Intercept)      Diethf
##   23.813333    3.020833
```

Examining the coefficients The following plot provides a visualization of the meaning of the coefficients with colored arrows (code not shown):

To make a connection with material presented earlier, this simple linear model is actually giving us the same result (the t-statistic and p-value) for the difference as a specific kind of t-test. This is the t-test between two groups with the assumption that the population standard deviation is the same for both groups. This was encoded into our linear model when we assumed that the errors ε were all equally distributed.

Although in this case the linear model is equivalent to a t-test, we will soon explore more complicated designs, where the linear model is a useful extension. Below we demonstrate that one does in fact get the exact same results:

Our lm estimates were:

```
summary(fit)$coefficients
```

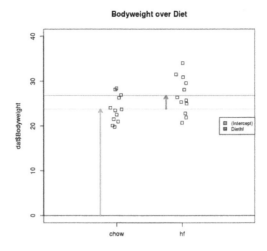

FIGURE 5.2
Estimated linear model coefficients for bodyweight data illustrated with arrows.

```
##               Estimate Std. Error    t value      Pr(>|t|)
## (Intercept) 23.813333   1.039353 22.911684 7.642256e-17
## Diethf       3.020833   1.469867  2.055174 5.192480e-02
```

And the t-statistic is the same:

```
ttest <- t.test(s[["hf"]], s[["chow"]], var.equal=TRUE)
summary(fit)$coefficients[2,3]
```

```
## [1] 2.055174
```

```
ttest$statistic
```

```
##        t
## 2.055174
```

5.5 Exercises

The function `lm` can be used to fit a simple, two group linear model. The test statistic from a linear model is equivalent to the test statistic we get when we perform a t-test with the equal variance assumption. Though the linear model in this case is equivalent to a t-test, we will soon explore more complicated designs, where the linear model is a useful extension (confounding variables, testing contrasts of terms, testing interactions, testing many terms at once, etc.).

Here we will review the mathematics on why these produce the same test statistic and therefore p-value.

We already know that the numerator of the t-statistic in both cases is the difference between the average of the groups, so we only have to see that the denominator is the same.

Of course, it makes sense that the denominator should be the same, since we are calculating the standard error of the same quantity (the difference) under the same assumptions (equal variance), but here we will show equivalence of the formula.

In the linear model, we saw how to calculate this standard error using the design matrix $\mathbf{X}$ and the estimate of σ^2 from the residuals. The estimate of σ^2 was the sum of squared residuals divided by $N - p$, where N is the total number of samples and p is the number of terms (an intercept and a group indicator, so here $p = 2$).

In the t-test, the denominator of the t-value is the standard error of the difference. The t-test formula for the standard error of the difference, if we assume equal variance in the two groups, is the square root of the variance:

$$\frac{1}{1/N_x + 1/N_y} \frac{\sum_{i=1}^{N_x}(X_i - \mu_x)^2 + \sum_{i=1}^{N_y}(Y_i - \mu_y)^2}{N_x + N_y - 2}$$

Here N_x is the number of samples in the first group and N_y is the number of samples in the second group.

If we look carefully, the second part of this equation is the sum of squared residuals, divided by $N - 2$.

All that is left to show is that the entry in the second row, second column of $(\mathbf{X}^\top\mathbf{X})^{-1}$ is $(1/N_x + 1/N_y)$

1. You can make a design matrix X for a two group comparison, either using `model.matrix` or simply with:

```
X <- cbind(rep(1,Nx + Ny),rep(c(0,1),c(Nx, Ny)))
```

In order to compare two groups, where the first group has Nx=5 samples and the second group has Ny=7 samples, what is the element in the 1st row and 1st column of $\mathbf{X}^\top\mathbf{X}$?

2. The other entries of $\mathbf{X}^\top\mathbf{X}$ are all the same. What is this number?

Now we just need to invert the matrix to obtain $(\mathbf{X}^\top\mathbf{X})^{-1}$. The formula for matrix inversion for a 2x2 matrix is as follows:

$$\begin{pmatrix} a & b \\ c & d \end{pmatrix}^{-1} = \frac{1}{ad - bc} \begin{pmatrix} d & -b \\ -c & a \end{pmatrix}$$

The element of the inverse in the 2nd row and the 2nd column is the element which will be used to calculate the standard error of the second coefficient of the linear model. This is $a/(ad - bc)$. And for our two group comparison, we saw that $a = N_x + N_y$ and the $b = c = d = N_y$. So it follows that this element is:

$$\frac{N_x + N_y}{(N_x + N_y)N_y - N_yN_y}$$

which simplifies to:

$$\frac{N_x + N_y}{N_xN_y} = 1/N_y + 1/N_x$$

5.6 Standard Errors

We have shown how to find the least squares estimates with matrix algebra. These estimates are random variables since they are linear combinations of the data. For these estimates to be useful, we also need to compute their standard errors. Linear algebra provides a powerful approach for this task. We provide several examples.

Falling object It is useful to think about where randomness comes from. In our falling object example, randomness was introduced through measurement errors. Each time we rerun the experiment, a new set of measurement errors will be made. This implies that our data will change randomly, which in turn suggests that our estimates will change randomly. For instance, our estimate of the gravitational constant will change every time we perform the experiment. The constant is fixed, but our estimates are not. To see this we can run a Monte Carlo simulation. Specifically, we will generate the data repeatedly and each time compute the estimate for the quadratic term.

```
set.seed(1)
B <- 10000
h0 <- 56.67
v0 <- 0
g <- 9.8 ##meters per second

n <- 25
tt <- seq(0,3.4,len=n) ##time in secs, t is a base function
X <-cbind(1,tt,tt^2)
##create X'X^-1 X'
A <- solve(crossprod(X)) %*% t(X)
betahat<-replicate(B,{
  y <- h0 + v0*tt  - 0.5*g*tt^2 + rnorm(n,sd=1)
  betahats <- A%*%y
  return(betahats[3])
})
head(betahat)
```

```
## [1] -5.038646 -4.894362 -5.143756 -5.220960 -5.063322 -4.777521
```

As expected, the estimate is different every time. This is because $\hat{\beta}$ is a random variable. It therefore has a distribution:

```
library(rafalib)
mypar(1,2)
hist(betahat)
qqnorm(betahat)
qqline(betahat)
```

Since $\hat{\beta}$ is a linear combination of the data which we made normal in our simulation, it is also normal as seen in the qq-plot above. Also, the mean of the distribution is the true parameter $-0.5g$, as confirmed by the Monte Carlo simulation performed above.

```
round(mean(betahat),1)
```

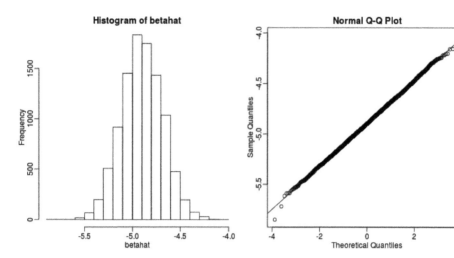

FIGURE 5.3
Distribution of estimated regression coefficients obtained from Monte Carlo simulated falling object data. The left is a histogram and on the right we have a qq-plot against normal theoretical quantiles.

```
## [1] -4.9
```

But we will not observe this exact value when we estimate because the standard error of our estimate is approximately:

```
sd(betahat)
```

```
## [1] 0.2129976
```

Here we will show how we can compute the standard error without a Monte Carlo simulation. Since in practice we do not know exactly how the errors are generated, we can't use the Monte Carlo approach.

Father and son heights In the father and son height examples, we have randomness because we have a random sample of father and son pairs. For the sake of illustration, let's assume that this is the entire population:

```
data(father.son,package="UsingR")
x <- father.son$fheight
y <- father.son$sheight
n <- length(y)
```

Now let's run a Monte Carlo simulation in which we take a sample size of 50 over and over again.

```
N <- 50
B <-1000
betahat <- replicate(B,{
  index <- sample(n,N)
  sampledat <- father.son[index,]
  x <- sampledat$fheight
```

```
  y <- sampledat$sheight
  lm(y~x)$coef
  })
betahat <- t(betahat) #have estimates in two columns
```

By making qq-plots, we see that our estimates are approximately normal random variables:

```
mypar(1,2)
qqnorm(betahat[,1])
qqline(betahat[,1])
qqnorm(betahat[,2])
qqline(betahat[,2])
```

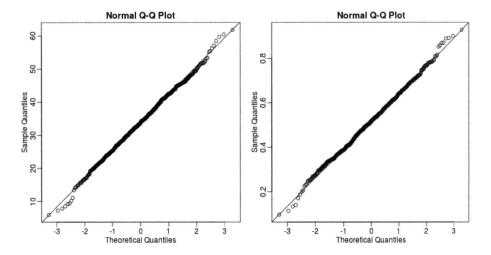

FIGURE 5.4
Distribution of estimated regression coefficients obtained from Monte Carlo simulated father-son height data. The left is a histogram and on the right we have a qq-plot against normal theoretical quantiles.

We also see that the correlation of our estimates is negative:

```
cor(betahat[,1],betahat[,2])
```

```
## [1] -0.9992293
```

When we compute linear combinations of our estimates, we will need to know this information to correctly calculate the standard error of these linear combinations.

In the next section, we will describe the variance-covariance matrix. The covariance of two random variables is defined as follows:

```
mean( (betahat[,1]-mean(betahat[,1] ))* (betahat[,2]-mean(betahat[,2])))
```

```
## [1] -1.035291
```

The covariance is the correlation multiplied by the standard deviations of each random variable:

$$\text{Corr}(X, Y) = \frac{\text{Cov}(X, Y)}{\sigma_X \sigma_Y}$$

Other than that, this quantity does not have a useful interpretation in practice. However, as we will see, it is a very useful quantity for mathematical derivations. In the next sections, we show useful matrix algebra calculations that can be used to estimate standard errors of linear model estimates.

Variance-covariance matrix (Advanced) As a first step we need to define the *variance-covariance matrix*, $\boldsymbol{\Sigma}$. For a vector of random variables, $\mathbf{Y}$, we define $\boldsymbol{\Sigma}$ as the matrix with the i, j entry:

$$\Sigma_{i,j} \equiv \text{Cov}(Y_i, Y_j)$$

The covariance is equal to the variance if $i = j$ and equal to 0 if the variables are independent. In the kinds of vectors considered up to now, for example, a vector $\mathbf{Y}$ of individual observations Y_i sampled from a population, we have assumed independence of each observation and assumed the Y_i all have the same variance σ^2, so the variance-covariance matrix has had only two kinds of elements:

$$\text{Cov}(Y_i, Y_i) = \text{var}(Y_i) = \sigma^2$$

$$\text{Cov}(Y_i, Y_j) = 0, \text{ for } i \neq j$$

which implies that $\boldsymbol{\Sigma} = \sigma^2 \mathbf{I}$ with $\mathbf{I}$, the identity matrix.

Later, we will see a case, specifically the estimate coefficients of a linear model, $\hat{\boldsymbol{\beta}}$, that has non-zero entries in the off diagonal elements of $\boldsymbol{\Sigma}$. Furthermore, the diagonal elements will not be equal to a single value σ^2.

Variance of a linear combination A useful result provided by linear algebra is that the variance covariance-matrix of a linear combination $\mathbf{AY}$ of $\mathbf{Y}$ can be computed as follows:

$$\text{var}(\mathbf{AY}) = \mathbf{A}\text{var}(\mathbf{Y})\mathbf{A}^\top$$

For example, if Y_1 and Y_2 are independent both with variance σ^2 then:

$$\text{var}\{Y_1 + Y_2\} = \text{var}\left\{(1 \quad 1)\begin{pmatrix} Y_1 \\ Y_2 \end{pmatrix}\right\}$$

$$= (1 \quad 1)\,\sigma^2\mathbf{I}\begin{pmatrix} 1 \\ 1 \end{pmatrix} = 2\sigma^2$$

as we expect. We use this result to obtain the standard errors of the LSE (least squares estimate).

LSE standard errors (Advanced) Note that $\hat{\boldsymbol{\beta}}$ is a linear combination of $\mathbf{Y}$: $\mathbf{AY}$ with $\mathbf{A} = (\mathbf{X}^\top\mathbf{X})^{-1}\mathbf{X}^\top$, so we can use the equation above to derive the variance of our estimates:

$$\text{var}(\hat{\boldsymbol{\beta}}) = \text{var}((\mathbf{X}^\top\mathbf{X})^{-1}\mathbf{X}^\top\mathbf{Y}) =$$

$$(\mathbf{X}^\top\mathbf{X})^{-1}\mathbf{X}^\top\text{var}(Y)((\mathbf{X}^\top\mathbf{X})^{-1}\mathbf{X}^\top)^\top =$$

$$(\mathbf{X}^\top\mathbf{X})^{-1}\mathbf{X}^\top\sigma^2\mathbf{I}((\mathbf{X}^\top\mathbf{X})^{-1}\mathbf{X}^\top)^\top =$$

$$\sigma^2(\mathbf{X}^\top\mathbf{X})^{-1}\mathbf{X}^\top\mathbf{X}(\mathbf{X}^\top\mathbf{X})^{-1} =$$

$$\sigma^2(\mathbf{X}^\top\mathbf{X})^{-1}$$

The diagonal of the square root of this matrix contains the standard error of our estimates.

Estimating σ^2 To obtain an actual estimate in practice from the formulas above, we need to estimate σ^2. Previously we estimated the standard errors from the sample. However, the sample standard deviation of Y is not σ because Y also includes variability introduced by the deterministic part of the model: $\mathbf{X}\beta$. The approach we take is to use the residuals.

We form the residuals like this:

$$\mathbf{r} \equiv \hat{\varepsilon} = \mathbf{Y} - \mathbf{X}\hat{\beta}$$

Both $\mathbf{r}$ and $\hat{\varepsilon}$ notations are used to denote residuals.

Then we use these to estimate, in a similar way, to what we do in the univariate case:

$$s^2 \equiv \hat{\sigma}^2 = \frac{1}{N-p}\mathbf{r}^\top\mathbf{r} = \frac{1}{N-p}\sum_{i=1}^{N}r_i^2$$

Here N is the sample size and p is the number of columns in $\mathbf{X}$ or number of parameters (including the intercept term β_0). The reason we divide by $N - p$ is because mathematical theory tells us that this will give us a better (unbiased) estimate.

Let's try this in R and see if we obtain the same values as we did with the Monte Carlo simulation above:

```
n <- nrow(father.son)
N <- 50
index <- sample(n,N)
sampledat <- father.son[index,]
x <- sampledat$fheight
y <- sampledat$sheight
X <- model.matrix(~x)

N <- nrow(X)
p <- ncol(X)

XtXinv <- solve(crossprod(X))

resid <- y - X %*% XtXinv %*% crossprod(X,y)

s <- sqrt( sum(resid^2)/(N-p))
ses <- sqrt(diag(XtXinv))*s
```

Let's compare to what `lm` provides:

```
summary(lm(y~x))$coef[,2]
```

```
## (Intercept)              x
##    8.3899781   0.1240767
```

```
ses
```

```
## (Intercept)              x
##    8.3899781   0.1240767
```

They are identical because they are doing the same thing. Also, note that we approximate the Monte Carlo results:

```
apply(betahat,2,sd)
```

```
## (Intercept)              x
##    8.3817556   0.1237362
```

Linear combination of estimates Frequently, we want to compute the standard deviation of a linear combination of estimates such as $\hat{\beta}_2 - \hat{\beta}_1$. This is a linear combination of $\hat{\beta}$:

$$\hat{\beta}_2 - \hat{\beta}_1 = \begin{pmatrix} 0 & -1 & 1 & 0 & \dots & 0 \end{pmatrix} \begin{pmatrix} \hat{\beta}_0 \\ \hat{\beta}_1 \\ \hat{\beta}_2 \\ \vdots \\ \hat{\beta}_p \end{pmatrix}$$

Using the above, we know how to compute the variance covariance matrix of $\hat{\beta}$.

CLT and t-distribution We have shown how we can obtain standard errors for our estimates. However, as we learned in the first chapter, to perform inference we need to know the distribution of these random variables. The reason we went through the effort to compute the standard errors is because the CLT applies in linear models. If N is large enough, then the LSE will be normally distributed with mean β and standard errors as described. For small samples, if the ε are normally distributed, then the $\hat{\beta} - \beta$ follow a t-distribution. We do not derive this result here, but the results are extremely useful since it is how we construct p-values and confidence intervals in the context of linear models.

Code versus math The standard approach to writing linear models either assume the values in $\mathbf{X}$ are fixed or that we are conditioning on them. Thus $\mathbf{X}\beta$ has no variance as the $\mathbf{X}$ is considered fixed. This is why we write $\text{var}(Y_i) = \text{var}(\varepsilon_i) = \sigma^2$. This can cause confusion in practice because if you, for example, compute the following:

```
x =  father.son$fheight
beta =  c(34,0.5)
var(beta[1]+beta[2]*x)
```

```
## [1] 1.883576
```

it is nowhere near 0. This is an example in which we have to be careful in distinguishing code from math. The function `var` is simply computing the variance of the list we feed it, while the mathematical definition of variance is considering only quantities that are random variables. In the R code above, `x` is not fixed at all: we are letting it vary, but when we write $\mathrm{var}(Y_i) = \sigma^2$ we are imposing, mathematically, x to be fixed. Similarly, if we use R to compute the variance of Y in our object dropping example, we obtain something very different than $\sigma^2 = 1$ (the known variance):

```
n <- length(tt)
y <- h0 + v0*tt  - 0.5*g*tt^2 + rnorm(n,sd=1)
var(y)
```

```
## [1] 329.5136
```

Again, this is because we are not fixing tt.

5.7 Exercises

In the previous assessment, we used a Monte Carlo technique to see that the linear model coefficients are random variables when the data is a random sample. Now we will use the previously seen matrix algebra to try to estimate the standard error of the linear model coefficients. Again, take a random sample of the **father.son** heights data:

```
library(UsingR)
N <- 50
set.seed(1)
index <- sample(n,N)
sampledat <- father.son[index,]
x <- sampledat$fheight
y <- sampledat$sheight
betahat <- lm(y~x)$coef
```

The formula for the standard error is:

$$\mathrm{SE}(\hat{\beta}) = \sqrt{\mathrm{var}(\hat{\beta})}$$

with:

$$\mathrm{var}(\hat{\beta}) = \sigma^2 (X^\top X)^{-1}$$

We will estimate or calculate each part of this equation and then combine them.

First, we want to estimate σ^2, the variance of Y. As we have seen in the previous unit, the random part of Y is only coming from ε, because we assume $X\beta$ is fixed. So we can try to estimate the variance of the ε's from the residuals, the Y_i minus the fitted values from the linear model.

1. The fitted values $\hat{Y}$ from a linear model can be obtained with:

```
fit <- lm(y ~ x)
fit$fitted.values
```

What is the sum of the squared residuals, where residuals are given by $r_i = Y_i - \hat{Y}_i$?

2. Our estimate of σ^2 will be the sum of squared residuals divided by $N - p$, the sample size minus the number of terms in the model. Since we have a sample of 50 and 2 terms in the model (an intercept and a slope), our estimate of σ^2 will be the sum of squared residuals divided by 48. Use the answer from exercise 1 to provide an estimate of σ^2.

3. Form the design matrix X (Note: use a capital X). This can be done by combining a column of 1's with a column containing x , the fathers' heights.

```
N <- 50
X <- cbind(rep(1,N), x)
```

Now calculate $(X^\top X)^{-1}$. Use the `solve` function for the inverse and `t` for the transpose. What is the element in the first row, first column?

4. Now we are one step away from the standard error of $\hat{\beta}$. Take the diagonals from the $(X^\top X)^{-1}$ matrix above, using the `diag` function. Multiply our estimate of σ^2 and the diagonals of this matrix. This is the estimated variance of $\hat{\beta}$, so take the square root of this. You should end up with two numbers: the standard error for the intercept and the standard error for the slope.

What is the standard error for the slope?

Compare your answer to this last question, to the value you estimated using Monte Carlo in the previous set of exercises. It will not be the same because we are only estimating the standard error given a particular sample of 50 (which we obtained with set.seed(1)).

Notice that the standard error estimate is also printed in the second column of:

```
summary(fit)
```

5.8   Interactions and Contrasts

As a running example to learn about more complex linear models, we will be using a dataset which compares the different frictional coefficients on the different legs of a spider. Specifically, we will be determining whether more friction comes from a pushing or pulling motion of the leg. The original paper from which the data was provided is:

Jonas O. Wolff & Stanislav N. Gorb, Radial arrangement of Janus-like setae permits friction control in spiders[1], Scientific Reports, 22 January 2013.

The abstract of the paper says,

The hunting spider Cupiennius salei (Arachnida, Ctenidae) possesses hairy attachment pads (claw tufts) at its distal legs, consisting of directional branched setae... Friction of claw tufts on smooth glass was measured to reveal the functional effect of seta arrangement within the pad.

Figure 1[2] includes some pretty cool electron microscope images of the tufts. We are interested in the comparisons in Figure 4[3], where the pulling and pushing motions are

[1] http://dx.doi.org/10.1038/srep01101
[2] http://www.nature.com/articles/srep01101/figures/1
[3] http://www.nature.com/articles/srep01101/figures/4

compared for different leg pairs (for a diagram of pushing and pulling see the top of Figure 3[4]).

We include the data in our dagdata package and can download it from here[5].

```
spider <- read.csv("spider_wolff_gorb_2013.csv", skip=1)
```

Initial visual inspection of the data Each measurement comes from one of our legs while it is either pushing or pulling. So we have two variables:

```
table(spider$leg,spider$type)
```

```
##
##      pull push
##  L1   34   34
##  L2   15   15
##  L3   52   52
##  L4   40   40
```

We can make a boxplot summarizing the measurements for each of the eight pairs. This is similar to Figure 4 of the original paper:

```
boxplot(spider$friction ~ spider$type * spider$leg,
        col=c("grey90","grey40"), las=2,
        main="Comparison of friction coefficients of different leg pairs")
```

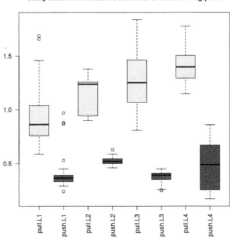

FIGURE 5.5
Comparison of friction coefficients of spiders' different leg pairs. The friction coefficient is calculated as the ratio of two forces (see paper Methods) so it is unitless.

What we can immediately see are two trends:

---

[4]http://www.nature.com/articles/srep01101/figures/3
[5]https://raw.githubusercontent.com/genomicsclass/dagdata/master/inst/extdata/spider_wolff_gorb_2013.csv

- The pulling motion has higher friction than the pushing motion.
- The leg pairs to the back of the spider (L4 being the last) have higher pulling friction.

Another thing to notice is that the groups have different spread around their average, what we call *within-group variance*. This is somewhat of a problem for the kinds of linear models we will explore below, since we will be assuming that around the population average values, the errors ε_i are distributed identically, meaning the same variance within each group. The consequence of ignoring the different variances for the different groups is that comparisons between those groups with small variances will be overly "conservative" (because the overall estimate of variance is larger than an estimate for just these groups), and comparisons between those groups with large variances will be overly confident. If the spread is related to the range of friction, such that groups with large friction values also have larger spread, a possibility is to transform the data with a function such as the `log` or `sqrt`. This looks like it could be useful here, since three of the four push groups (L1, L2, L3) have the smallest friction values and also the smallest spread.

Some alternative tests for comparing groups without transforming the values first include: t-tests without the equal variance assumption using a "Welch" or "Satterthwaite approximation", or the Wilcoxon rank sum test mentioned previously. However here, for simplicity of illustration, we will fit a model that assumes equal variance and shows the different kinds of linear model designs using this dataset, setting aside the issue of different within-group variances.

A linear model with one variable To remind ourselves how the simple two-group linear model looks, we will subset the data to include only the L1 leg pair, and run `lm`:

```
spider.sub <- spider[spider$leg == "L1",]
fit <- lm(friction ~ type, data=spider.sub)
summary(fit)
```

```
##
## Call:
## lm(formula = friction ~ type, data = spider.sub)
##
## Residuals:
##      Min       1Q   Median       3Q      Max
## -0.33147 -0.10735 -0.04941 -0.00147  0.76853
##
## Coefficients:
##             Estimate Std. Error t value Pr(>|t|)
## (Intercept)  0.92147    0.03827  24.078  < 2e-16 ***
## typepush    -0.51412    0.05412  -9.499  5.7e-14 ***
## ---
## Signif. codes:  0 '***' 0.001 '**' 0.01 '*' 0.05 '.' 0.1 ' ' 1
##
## Residual standard error: 0.2232 on 66 degrees of freedom
## Multiple R-squared:  0.5776, Adjusted R-squared:  0.5711
## F-statistic: 90.23 on 1 and 66 DF,  p-value: 5.698e-14
```

```
(coefs <- coef(fit))
```

```
## (Intercept)    typepush
##   0.9214706  -0.5141176
```

These two estimated coefficients are the mean of the pull observations (the first estimated coefficient) and the difference between the means of the two groups (the second coefficient). We can show this with R code:

```
s <- split(spider.sub$friction, spider.sub$type)
mean(s[["pull"]])
```

```
## [1] 0.9214706
```

```
mean(s[["push"]]) - mean(s[["pull"]])
```

```
## [1] -0.5141176
```

We can form the design matrix, which was used inside `lm`:

```
X <- model.matrix(~ type, data=spider.sub)
colnames(X)
```

```
## [1] "(Intercept)" "typepush"
```

```
head(X)
```

```
##   (Intercept) typepush
## 1           1        0
## 2           1        0
## 3           1        0
## 4           1        0
## 5           1        0
## 6           1        0
```

```
tail(X)
```

```
##    (Intercept) typepush
## 63           1        1
## 64           1        1
## 65           1        1
## 66           1        1
## 67           1        1
## 68           1        1
```

Now we'll make a plot of the $\mathbf{X}$ matrix by putting a black block for the 1's and a white block for the 0's. This plot will be more interesting for the linear models later on in this script. Along the y-axis is the sample number (the row number of the `data`) and along the x-axis is the column of the design matrix $\mathbf{X}$. If you have installed the *rafalib* library, you can make this plot with the `imagemat` function:

```
library(rafalib)
imagemat(X, main="Model matrix for linear model with one variable")
```

Examining the estimated coefficients Now we show the coefficient estimates from the linear model in a diagram with arrows (code not shown).

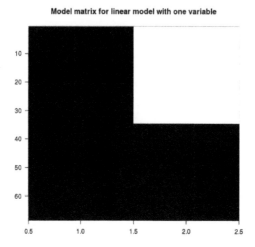

FIGURE 5.6
Model matrix for linear model with one variable.

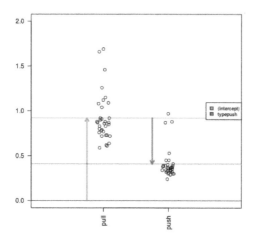

FIGURE 5.7
Diagram of the estimated coefficients in the linear model. The green arrow indicates the Intercept term, which goes from zero to the mean of the reference group (here the 'pull' samples). The orange arrow indicates the difference between the push group and the pull group, which is negative in this example. The circles show the individual samples, jittered horizontally to avoid overplotting.

A linear model with two variables Now we'll continue and examine the full dataset, including the observations from all leg pairs. In order to model both the leg pair differences (L1, L2, L3, L4) and the push vs. pull difference, we need to include both terms in the R formula. Let's see what kind of design matrix will be formed with two variables in the formula:

```
X <- model.matrix(~ type + leg, data=spider)
colnames(X)

## [1] "(Intercept)" "typepush"    "legL2"       "legL3"       "legL4"

head(X)

##   (Intercept) typepush legL2 legL3 legL4
## 1           1        0     0     0     0
## 2           1        0     0     0     0
## 3           1        0     0     0     0
## 4           1        0     0     0     0
## 5           1        0     0     0     0
## 6           1        0     0     0     0

imagemat(X, main="Model matrix for linear model with two factors")
```

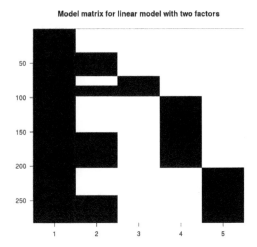

FIGURE 5.8
Image of the model matrix for a formula with type + leg

The first column is the intercept, and so it has 1's for all samples. The second column has 1's for the push samples, and we can see that there are four groups of them. Finally, the third, fourth and fifth columns have 1's for the L2, L3 and L4 samples. The L1 samples do not have a column, because *L1* is the reference level for `leg`. Similarly, there is no *pull* column, because *pull* is the reference level for the `type` variable.

To estimate coefficients for this model, we use `lm` with the formula ~ `type + leg`. We'll save the linear model to `fitTL` standing for a *fit* with *Type* and *Leg*.

```
fitTL <- lm(friction ~ type + leg, data=spider)
summary(fitTL)
```

```
##
## Call:
## lm(formula = friction ~ type + leg, data = spider)
##
## Residuals:
##      Min       1Q   Median       3Q      Max
## -0.46392 -0.13441 -0.00525  0.10547  0.69509
##
## Coefficients:
##              Estimate Std. Error t value Pr(>|t|)
## (Intercept)   1.05392    0.02816  37.426  < 2e-16 ***
## typepush     -0.77901    0.02482 -31.380  < 2e-16 ***
## legL2         0.17192    0.04569   3.763 0.000205 ***
## legL3         0.16049    0.03251   4.937 1.37e-06 ***
## legL4         0.28134    0.03438   8.183 1.01e-14 ***
## ---
## Signif. codes:  0 '***' 0.001 '**' 0.01 '*' 0.05 '.' 0.1 ' ' 1
##
## Residual standard error: 0.2084 on 277 degrees of freedom
## Multiple R-squared:  0.7916, Adjusted R-squared:  0.7886
## F-statistic:   263 on 4 and 277 DF,  p-value: < 2.2e-16
```

```
(coefs <- coef(fitTL))
```

```
## (Intercept)    typepush       legL2       legL3       legL4
##   1.0539153  -0.7790071   0.1719216   0.1604921   0.2813382
```

R uses the name **coefficient** to denote the component containing the least squares **estimates**. It is important to remember that the coefficients are parameters that we do not observe, but only estimate.

Mathematical representation The model we are fitting above can be written as

$$Y_i = \beta_0 + \beta_1 x_{i,1} + \beta_2 x_{i,2} + \beta_3 x_{i,3} + \beta_4 x_{i,4} + \varepsilon_i, i = 1, \ldots, N$$

with the x all indicator variables denoting push or pull and which leg. For example, a push on leg 3 will have $x_{i,1}$ and $x_{i,3}$ equal to 1 and the rest would be 0. Throughout this section we will refer to the β s with the effects they represent. For example we call β_0 the intercept, β_1 the pull effect, β_2 the L2 effect, etc. We do not observe the coefficients, e.g. β_1, directly, but estimate them with, e.g. $\hat{\beta}_4$.

We can now form the matrix $\mathbf{X}$ depicted above and obtain the least square estimates with:

$$\hat{\beta} = (\mathbf{X}^\top \mathbf{X})^{-1} \mathbf{X}^\top \mathbf{Y}$$

```
Y <- spider$friction
X <- model.matrix(~ type + leg, data=spider)
beta.hat <- solve(t(X) %*% X) %*% t(X) %*% Y
t(beta.hat)
```

```
##        (Intercept)    typepush       legL2       legL3       legL4
## [1,]     1.053915  -0.7790071  0.1719216  0.1604921  0.2813382
```

```
coefs
```

```
## (Intercept)      typepush        legL2        legL3        legL4
##    1.0539153   -0.7790071    0.1719216    0.1604921    0.2813382
```

We can see that these values agree with the output of `lm`.

Examining the estimated coefficients We can make the same plot as before, with arrows for each of the estimated coefficients in the model (code not shown).

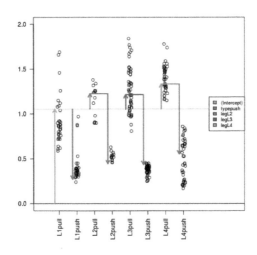

FIGURE 5.9
Diagram of the estimated coefficients in the linear model. As before, the teal-green arrow represents the Intercept, which fits the mean of the reference group (here, the pull samples for leg L1). The purple, pink, and yellow-green arrows represent differences between the three other leg groups and L1. The orange arrow represents the difference between the push and pull samples for all groups.

In this case, the fitted means for each group, derived from the fitted coefficients, do not line up with those we obtain from simply taking the average from each of the eight possible groups. The reason is that our model uses five coefficients, instead of eight. We are **assuming** that the effects are additive. However, as we demonstrate in more detail below, this particular dataset is better described with a model including interactions.

```
s <- split(spider$friction, spider$group)
mean(s[["L1pull"]])
```

```
## [1] 0.9214706
```

```
coefs[1]
```

```
## (Intercept)
##    1.053915
```

```
mean(s[["L1push"]])
```

```
## [1] 0.4073529
```

```
coefs[1] + coefs[2]
```

```
## (Intercept)
##   0.2749082
```

Here we can demonstrate that the push vs. pull estimated coefficient, `coefs[2]`, is a weighted average of the difference of the means for each group. Furthermore, the weighting is determined by the sample size of each group. The math works out simply here because the sample size is equal for the push and pull subgroups within each leg pair. If the sample sizes were not equal for push and pull within each leg pair, the weighting is more complicated but uniquely determined by a formula involving the sample size of each subgroup, the total sample size, and the number of coefficients. This can be worked out from $(\mathbf{X}^\top \mathbf{X})^{-1} \mathbf{X}^\top$.

```
means <- sapply(s, mean)
##the sample size of push or pull groups for each leg pair
ns <- sapply(s, length)[c(1,3,5,7)]
(w <- ns/sum(ns))
```

```
##    L1pull    L2pull    L3pull    L4pull
## 0.2411348 0.1063830 0.3687943 0.2836879
```

```
sum(w * (means[c(2,4,6,8)] - means[c(1,3,5,7)]))
```

```
## [1] -0.7790071
```

```
coefs[2]
```

```
##    typepush
## -0.7790071
```

Contrasting coefficients Sometimes, the comparison we are interested in is represented directly by a single coefficient in the model, such as the push vs. pull difference, which was `coefs[2]` above. However, sometimes, we want to make a comparison which is not a single coefficient, but a combination of coefficients, which is called a *contrast*. To introduce the concept of *contrasts*, first consider the comparisons which we can read off from the linear model summary:

```
coefs
```

```
## (Intercept)    typepush       legL2       legL3       legL4
##   1.0539153  -0.7790071   0.1719216   0.1604921   0.2813382
```

Here we have the intercept estimate, the push vs. pull estimated effect across all leg pairs, and the estimates for the L2 vs. L1 effect, the L3 vs. L1 effect, and the L4 vs. L1 effect. What if we want to compare two groups and one of those groups is not L1? The solution to this question is to use *contrasts*.

A *contrast* is a combination of estimated coefficient: $\mathbf{c}^\top \hat{\boldsymbol{\beta}}$, where $\mathbf{c}$ is a column vector with as many rows as the number of coefficients in the linear model. If $\mathbf{c}$ has a 0 for one or more of its rows, then the corresponding estimated coefficients in $\hat{\boldsymbol{\beta}}$ are not involved in the contrast.

If we want to compare leg pairs L3 and L2, this is equivalent to contrasting two coefficients from the linear model because, in this contrast, the comparison to the reference level *L1* cancels out:

$$(L3 - L1) - (L2 - L1) = L3 - L2$$

An easy way to make these contrasts of two groups is to use the `contrast` function from the *contrast* package. We just need to specify which groups we want to compare. We have to pick one of *pull* or *push* types, although the answer will not differ, as we will see below.

```
library(contrast) #Available from CRAN
L3vsL2 <- contrast(fitTL,list(leg="L3",type="pull"),list(leg="L2",type="pull"))
L3vsL2
```

```
## lm model parameter contrast
##
##     Contrast       S.E.      Lower       Upper     t  df Pr(>|t|)
##  -0.01142949 0.04319685 -0.0964653 0.07360632 -0.26 277   0.7915
```

The first column `Contrast` gives the L3 vs. L2 estimate from the model we fit above.

We can show that the least squares estimates of a linear combination of coefficients is the same linear combination of the estimates. Therefore, the effect size estimate is just the difference between two estimated coefficients. The contrast vector used by `contrast` is stored as a variable called `X` within the resulting object (not to be confused with our original **X**, the design matrix).

```
coefs[4] - coefs[3]
```

```
##       legL3
## -0.01142949
```

```
(cT <- L3vsL2$X)
```

```
##   (Intercept) typepush legL2 legL3 legL4
## 1           0        0    -1     1     0
## attr(,"assign")
## [1] 0 1 2 2 2
## attr(,"contrasts")
## attr(,"contrasts")$type
## [1] "contr.treatment"
##
## attr(,"contrasts")$leg
## [1] "contr.treatment"
```

```
cT %*% coefs
```

```
##             [,1]
## 1 -0.01142949
```

What about the standard error and t-statistic? As before, the t-statistic is the estimate divided by the standard error. The standard error of the contrast estimate is formed by multiplying the contrast vector **c** on either side of the estimated covariance matrix, $\hat{\Sigma}$, our estimate for $\mathrm{var}(\hat{\beta})$:

$$\sqrt{\mathbf{c}^\top \hat{\boldsymbol{\Sigma}} \mathbf{c}}$$

where we saw the covariance of the coefficients earlier:

$$\boldsymbol{\Sigma} = \sigma^2 (\mathbf{X}^\top \mathbf{X})^{-1}$$

We estimate σ^2 with the sample estimate $\hat{\sigma}^2$ described above and obtain:

```
Sigma.hat <- sum(fitTL$residuals^2)/(nrow(X) - ncol(X)) * solve(t(X) %*% X)
signif(Sigma.hat, 2)
```

```
##              (Intercept) typepush     legL2     legL3     legL4
## (Intercept)      0.00079 -3.1e-04  -0.00064  -0.00064  -0.00064
## typepush        -0.00031  6.2e-04   0.00000   0.00000   0.00000
## legL2           -0.00064 -6.4e-20   0.00210   0.00064   0.00064
## legL3           -0.00064 -6.4e-20   0.00064   0.00110   0.00064
## legL4           -0.00064 -1.2e-19   0.00064   0.00064   0.00120
```

```
sqrt(cT %*% Sigma.hat %*% t(cT))
```

```
##            1
## 1 0.04319685
```

```
L3vsL2$SE
```

```
## [1] 0.04319685
```

We would have obtained the same result for a contrast of L3 and L2 had we picked `type="push"`. The reason it does not change the contrast is because it leads to addition of the `typepush` effect on both sides of the difference, which cancels out:

```
L3vsL2.equiv <- contrast(fitTL,list(leg="L3",type="push"),list(leg="L2",type="push"))
L3vsL2.equiv$X
```

```
##   (Intercept) typepush legL2 legL3 legL4
## 1           0        0    -1     1     0
## attr(,"assign")
## [1] 0 1 2 2 2
## attr(,"contrasts")
## attr(,"contrasts")$type
## [1] "contr.treatment"
##
## attr(,"contrasts")$leg
## [1] "contr.treatment"
```

5.9   Linear Model with Interactions

In the previous linear model, we assumed that the push vs. pull effect was the same for all of the leg pairs (the same orange arrow). You can easily see that this does not capture the trends in the data that well. That is, the tips of the arrows did not line up perfectly with the group averages. For the L1 leg pair, the push vs. pull estimated coefficient was too large, and for the L3 leg pair, the push vs. pull coefficient was somewhat too small.

Interaction terms will help us overcome this problem by introducing additional coefficients to compensate for differences in the push vs. pull effect across the 4 groups. As we already have a push vs. pull term in the model, we only need to add three more terms to have the freedom to find leg-pair-specific push vs. pull differences. As we will see, interaction terms are added to the design matrix by multiplying the columns of the design matrix representing existing terms.

We can rebuild our linear model with an interaction between `type` and `leg`, by including an extra term in the formula `type:leg`. The : symbol adds an interaction between the two variables surrounding it. An equivalent way to specify this model is `~ type*leg`, which will expand to the formula `~ type + leg + type:leg`, with main effects for `type`, `leg` and an interaction term `type:leg`.

```
X <- model.matrix(~ type + leg + type:leg, data=spider)
colnames(X)
```

```
## [1] "(Intercept)"      "typepush"        "legL2"           "legL3"
## [5] "legL4"            "typepush:legL2"  "typepush:legL3"  "typepush:legL4"
```

```
head(X)
```

```
##   (Intercept) typepush legL2 legL3 legL4 typepush:legL2 typepush:legL3
## 1           1        0     0     0     0              0              0
## 2           1        0     0     0     0              0              0
## 3           1        0     0     0     0              0              0
## 4           1        0     0     0     0              0              0
## 5           1        0     0     0     0              0              0
## 6           1        0     0     0     0              0              0
##   typepush:legL4
## 1              0
## 2              0
## 3              0
## 4              0
## 5              0
## 6              0
```

```
imagemat(X, main="Model matrix for linear model with interactions")
```

Columns 6-8 (`typepush:legL2`, `typepush:legL3`, and `typepush:legL4`) are the product of the 2nd column (`typepush`) and columns 3-5 (the three `leg` columns). Looking at the last column, for example, the `typepush:legL4` column is adding an extra coefficient $\beta_{push,L4}$ to those samples which are both push samples and leg pair L4 samples. This accounts for a possible difference when the mean of samples in the L4-push group are not at the location which would be predicted by adding the estimated intercept, the estimated push coefficient `typepush`, and the estimated L4 coefficient `legL4`.

We can run the linear model using the same code as before:

```
fitX <- lm(friction ~ type + leg + type:leg, data=spider)
summary(fitX)
```

```
##
## Call:
```

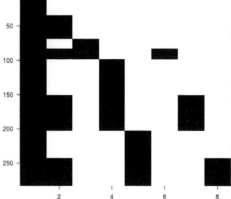

FIGURE 5.10
Image of model matrix with interactions.

```
## lm(formula = friction ~ type + leg + type:leg, data = spider)
##
## Residuals:
##      Min       1Q   Median       3Q      Max
## -0.46385 -0.10735 -0.01111  0.07848  0.76853
##
## Coefficients:
##                  Estimate Std. Error t value Pr(>|t|)
## (Intercept)       0.92147    0.03266  28.215  < 2e-16 ***
## typepush         -0.51412    0.04619 -11.131  < 2e-16 ***
## legL2             0.22386    0.05903   3.792 0.000184 ***
## legL3             0.35238    0.04200   8.390 2.62e-15 ***
## legL4             0.47928    0.04442  10.789  < 2e-16 ***
## typepush:legL2   -0.10388    0.08348  -1.244 0.214409
## typepush:legL3   -0.38377    0.05940  -6.461 4.73e-10 ***
## typepush:legL4   -0.39588    0.06282  -6.302 1.17e-09 ***
## ---
## Signif. codes:  0 '***' 0.001 '**' 0.01 '*' 0.05 '.' 0.1 ' ' 1
##
## Residual standard error: 0.1904 on 274 degrees of freedom
## Multiple R-squared:  0.8279, Adjusted R-squared:  0.8235
## F-statistic: 188.3 on 7 and 274 DF,  p-value: < 2.2e-16

coefs <- coef(fitX)
```

Examining the estimated coefficients Here is where the plot with arrows really helps us interpret the coefficients. The estimated interaction coefficients (the yellow, brown and silver arrows) allow leg-pair-specific differences in the push vs. pull difference. The orange arrow now represents the estimated push vs. pull difference only for the reference leg pair, which is L1. If an estimated interaction coefficient is large, this means that the push vs. pull difference for that leg pair is very different than the push vs. pull difference in the reference leg pair.

Now, as we have eight terms in the model and eight parameters, you can check that the tips of the arrowheads are exactly equal to the group means (code not shown).

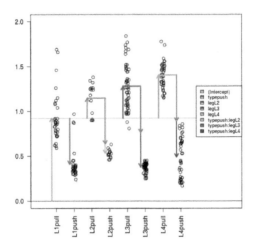

FIGURE 5.11
Diagram of the estimated coefficients in the linear model. In the design with interaction terms, the orange arrow now indicates the push vs. pull difference only for the reference group (L1), while three new arrows (yellow, brown and grey) indicate the additional push vs. pull differences in the non-reference groups (L2, L3 and L4) with respect to the reference group.

Contrasts Again we will show how to combine estimated coefficients from the model using contrasts. For some simple cases, we can use the contrast package. Suppose we want to know the push vs. pull effect for the L2 leg pair samples. We can see from the arrow plot that this is the orange arrow plus the yellow arrow. We can also specify this comparison with the `contrast` function:

```
library(contrast) ##Available from CRAN
L2push.vs.pull <- contrast(fitX,
                list(leg="L2", type = "push"),
                list(leg="L2", type = "pull"))
L2push.vs.pull

## lm model parameter contrast
##
## Contrast      S.E.      Lower        Upper      t  df Pr(>|t|)
##    -0.618 0.0695372 -0.7548951 -0.4811049 -8.89 274        0

coefs[2] + coefs[6] ##we know this is also orange + yellow arrow

## typepush
##    -0.618
```

Differences of differences The question of whether the push vs. pull difference is *different* in L2 compared to L1, is answered by a single term in the model: the `typepush:legL2` estimated coefficient corresponding to the yellow arrow in the plot. A p-value for whether this coefficient is actually equal to zero can be read off from the table printed with `summary(fitX)` above. Similarly, we can read off the p-values for the differences of differences for L3 vs. L1 and for L4 vs. L1.

Suppose we want to know if the push vs. pull difference is *different* in L3 compared to L2. By examining the arrows in the diagram above, we can see that the push vs. pull effect for a leg pair other than L1 is the `typepush` arrow plus the interaction term for that group.

If we work out the math for comparing across two non-reference leg pairs, this is:

$$(\text{typepush} + \text{typepush:legL3}) - (\text{typepush} + \text{typepush:legL2})$$

...which simplifies to:

$$= \text{typepush:legL3} - \text{typepush:legL2}$$

We can't make this contrast using the `contrast` function shown before, but we can make this comparison using the `glht` (for "general linear hypothesis test") function from the *multcomp* package. We need to form a 1-row matrix which has a -1 for the `typepush:legL2` coefficient and a +1 for the `typepush:legL3` coefficient. We provide this matrix to the `linfct` (linear function) argument, and obtain a summary table for this contrast of estimated interaction coefficients.

Note that there are other ways to perform contrasts using base R, and this is just our preferred way.

```
library(multcomp) ##Available from CRAN
C <- matrix(c(0,0,0,0,0,-1,1,0), 1)
L3vsL2interaction <- glht(fitX, linfct=C)
summary(L3vsL2interaction)

##
##    Simultaneous Tests for General Linear Hypotheses
##
## Fit: lm(formula = friction ~ type + leg + type:leg, data = spider)
##
## Linear Hypotheses:
##         Estimate Std. Error t value Pr(>|t|)
## 1 == 0 -0.27988    0.07893  -3.546  0.00046 ***
## ---
## Signif. codes:  0 '***' 0.001 '**' 0.01 '*' 0.05 '.' 0.1 ' ' 1
## (Adjusted p values reported -- single-step method)

coefs[7] - coefs[6] ##we know this is also brown - yellow

## typepush:legL3
##     -0.2798846
```

5.10   Analysis of Variance

Suppose that we want to know if the push vs. pull difference is different across leg pairs in general. We do not want to compare any two leg pairs in particular, but rather we want to know if the three interaction terms which represent differences in the push vs. pull difference across leg pairs are larger than we would expect them to be if the push vs. pull difference was in fact equal across all leg pairs.

Such a question can be answered by an *analysis of variance*, which is often abbreviated as ANOVA. ANOVA compares the reduction in the sum of squares of the residuals for models of different complexity. The model with eight coefficients is more complex than the model with five coefficients where we assumed the push vs. pull difference was equal across leg pairs. The least complex model would only use a single coefficient, an intercept. Under certain assumptions we can also perform inference that determines the probability of improvements as large as what we observed. Let's first print the result of an ANOVA in R and then examine the results in detail:

```
anova(fitX)

## Analysis of Variance Table
##
## Response: friction
##               Df Sum Sq Mean Sq  F value    Pr(>F)
## type           1 42.783  42.783 1179.713 < 2.2e-16 ***
## leg            3  2.921   0.974   26.847 2.972e-15 ***
## type:leg       3  2.098   0.699   19.282 2.256e-11 ***
## Residuals    274  9.937   0.036
## ---
## Signif. codes:  0 '***' 0.001 '**' 0.01 '*' 0.05 '.' 0.1 ' ' 1
```

The first line tells us that adding a variable `type` (push or pull) to the design is very useful (reduces the sum of squared residuals) compared to a model with only an intercept. We can see that it is useful, because this single coefficient reduces the sum of squares by 42.783. The original sum of squares of the model with just an intercept is:

```
mu0 <- mean(spider$friction)
(initial.ss <- sum((spider$friction - mu0)^2))

## [1] 57.73858
```

Note that this initial sum of squares is just a scaled version of the sample variance:

```
N <- nrow(spider)
(N - 1) * var(spider$friction)

## [1] 57.73858
```

Let's see exactly how we get this 42.783. We need to calculate the sum of squared residuals for the model with only the type information. We can do this by calculating the residuals, squaring these, summing these within groups and then summing across the groups.

```
s <- split(spider$friction, spider$type)
after.type.ss <- sum( sapply(s, function(x) {
  residual <- x - mean(x)
  sum(residual^2)
}) )
```

The reduction in sum of squared residuals from introducing the `type` coefficient is therefore:

```
(type.ss <- initial.ss - after.type.ss)
```

```
## [1] 42.78307
```

Through simple arithmetic[6], this reduction can be shown to be equivalent to the sum of squared differences between the fitted values for the models with formula `~type` and `~1`:

```
sum(sapply(s, length) * (sapply(s, mean) - mu0)^2)
```

```
## [1] 42.78307
```

Keep in mind that the order of terms in the formula, and therefore rows in the ANOVA table, is important: each row considers the reduction in the sum of squared residuals after adding coefficients *compared to the model in the previous row*.

The other columns in the ANOVA table show the "degrees of freedom" with each row. As the `type` variable introduced only one term in the model, the `Df` column has a 1. Because the `leg` variable introduced three terms in the model (`legL2`, `legL3` and `legL4`), the `Df` column has a 3.

Finally, there is a column which lists the *F value*. The F value is the *mean of squares* for the inclusion of the terms of interest (the sum of squares divided by the degrees of freedom) divided by the mean squared residuals (from the bottom row):

$$r_i = Y_i - \hat{Y}_i$$

$$\text{Mean Sq Residuals} = \frac{1}{N-p} \sum_{i=1}^{N} r_i^2$$

where p is the number of coefficients in the model (here eight, including the intercept term).

Under the null hypothesis (the true value of the additional coefficient(s) is 0), we have a theoretical result for what the distribution of the F value will be for each row. The assumptions needed for this approximation to hold are similar to those of the t-distribution approximation we described in earlier chapters. We either need a large sample size so that CLT applies or we need the population data to follow a normal approximation.

As an example of how one interprets these p-values, let's take the last row `type:leg` which specifies the three interaction coefficients. Under the null hypothesis that the true value for these three additional terms is actually 0, e.g. $\beta_{\text{push,L2}} = 0, \beta_{\text{push,L3}} = 0, \beta_{\text{push,L4}} = 0$, then we can calculate the chance of seeing such a large F-value for this row of the ANOVA table. Remember that we are only concerned with large values here, because we have a ratio of sum of squares, the F-value can only be positive. The p-value in the last column for the `type:leg` row can be interpreted as: under the null hypothesis that there are no differences in the push vs. pull difference across leg pair, this is the probability of an estimated interaction coefficient explaining so much of the observed variance. If this p-value is small, we would consider rejecting the null hypothesis that the push vs. pull difference is the same across leg pairs.

The F distribution[7] has two parameters: one for the degrees of freedom of the numerator (the terms of interest) and one for the denominator (the residuals). In the case of the interaction coefficients row, this is 3, the number of interaction coefficients divided by 274, the number of samples minus the total number of coefficients.

---

[6] http://en.wikipedia.org/wiki/Partition_of_sums_of_squares#Proof
[7] http://en.wikipedia.org/wiki/F-distribution

A different specification of the same model Now we show an alternate specification of the same model, wherein we assume that each combination of type and leg has its own mean value (and so that the push vs. pull effect is not the same for each leg pair). This specification is in some ways simpler, as we will see, but it does not allow us to build the ANOVA table as above, because it does not split interaction coefficients out in the same way.

We start by constructing a factor variable with a level for each unique combination of type and leg. We include a 0 + in the formula because we do not want to include an intercept in the model matrix.

```
##earlier, we defined the 'group' column:
spider$group <- factor(paste0(spider$leg, spider$type))
X <- model.matrix(~ 0 + group, data=spider)
colnames(X)
```

```
## [1] "groupL1pull" "groupL1push" "groupL2pull" "groupL2push" "groupL3pull"
## [6] "groupL3push" "groupL4pull" "groupL4push"
```

```
head(X)
```

```
##   groupL1pull groupL1push groupL2pull groupL2push groupL3pull groupL3push
## 1           1           0           0           0           0           0
## 2           1           0           0           0           0           0
## 3           1           0           0           0           0           0
## 4           1           0           0           0           0           0
## 5           1           0           0           0           0           0
## 6           1           0           0           0           0           0
##   groupL4pull groupL4push
## 1           0           0
## 2           0           0
## 3           0           0
## 4           0           0
## 5           0           0
## 6           0           0
```

```
imagemat(X, main="Model matrix for linear model with group variable")
```

We can run the linear model with the familiar call:

```
fitG <- lm(friction ~ 0 + group, data=spider)
summary(fitG)
```

```
##
## Call:
## lm(formula = friction ~ 0 + group, data = spider)
##
## Residuals:
##      Min       1Q   Median       3Q      Max
## -0.46385 -0.10735 -0.01111  0.07848  0.76853
##
## Coefficients:
##             Estimate Std. Error t value Pr(>|t|)
## groupL1pull  0.92147    0.03266   28.21   <2e-16 ***
```

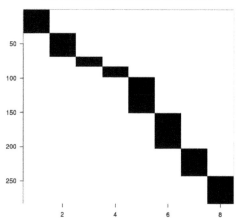

FIGURE 5.12

Image of model matrix for linear model with group variable. This model, also with eight terms, gives a unique fitted value for each combination of type and leg.

```
## groupL1push  0.40735    0.03266   12.47   <2e-16 ***
## groupL2pull  1.14533    0.04917   23.29   <2e-16 ***
## groupL2push  0.52733    0.04917   10.72   <2e-16 ***
## groupL3pull  1.27385    0.02641   48.24   <2e-16 ***
## groupL3push  0.37596    0.02641   14.24   <2e-16 ***
## groupL4pull  1.40075    0.03011   46.52   <2e-16 ***
## groupL4push  0.49075    0.03011   16.30   <2e-16 ***
## ---
## Signif. codes:  0 '***' 0.001 '**' 0.01 '*' 0.05 '.' 0.1 ' ' 1
##
## Residual standard error: 0.1904 on 274 degrees of freedom
## Multiple R-squared:  0.96,  Adjusted R-squared:  0.9588
## F-statistic:   821 on 8 and 274 DF,  p-value: < 2.2e-16

coefs <- coef(fitG)
```

Examining the estimated coefficients Now we have eight arrows, one for each group. The arrow tips align directly with the mean of each group:

Simple contrasts using the contrast package While we cannot perform an ANOVA with this formulation, we can easily contrast the estimated coefficients for individual groups using the contrast function:

```
groupL2push.vs.pull <- contrast(fitG,
                          list(group = "L2push"),
                          list(group = "L2pull"))
groupL2push.vs.pull

## lm model parameter contrast
##
```

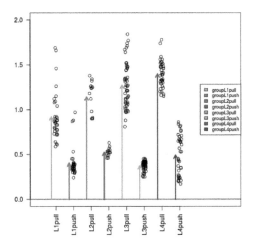

FIGURE 5.13
Diagram of the estimated coefficients in the linear model, with each term representing the mean of a combination of type and leg.

```
##   Contrast      S.E.       Lower       Upper      t  df Pr(>|t|)
## 1   -0.618 0.0695372 -0.7548951 -0.4811049 -8.89 274        0
```

```
coefs[4] - coefs[3]
```

```
## groupL2push
##      -0.618
```

Differences of differences when there is no intercept We can also make pair-wise comparisons of the estimated push vs. pull difference across leg pair. For example, if we want to compare the push vs. pull difference in leg pair L3 vs. leg pair L2:

$$(L3push - L3pull) - (L2push - L2pull)$$

$$= L3\ push + L2pull - L3pull - L2push$$

```
C <- matrix(c(0,0,1,-1,-1,1,0,0), 1)
groupL3vsL2interaction <- glht(fitG, linfct=C)
summary(groupL3vsL2interaction)
```

```
##
##   Simultaneous Tests for General Linear Hypotheses
##
## Fit: lm(formula = friction ~ 0 + group, data = spider)
##
## Linear Hypotheses:
##        Estimate Std. Error t value Pr(>|t|)
## 1 == 0 -0.27988    0.07893  -3.546  0.00046 ***
## ---
```

```
## Signif. codes:   0 '***' 0.001 '**' 0.01 '*' 0.05 '.' 0.1 ' ' 1
## (Adjusted p values reported -- single-step method)

names(coefs)

## [1] "groupL1pull" "groupL1push" "groupL2pull" "groupL2push" "groupL3pull"
## [6] "groupL3push" "groupL4pull" "groupL4push"

(coefs[6] - coefs[5]) - (coefs[4] - coefs[3])

## groupL3push
##  -0.2798846
```

5.11   Exercises

Suppose we have an experiment with two species A and B, and two conditions, control and treated.

```
species <- factor(c("A","A","B","B"))
condition <- factor(c("control","treated","control","treated"))
```

We will use the formula of ~ species + condition to create the model matrix:

```
model.matrix(~ species + condition)
```

1. Suppose we want to build a contrast of coefficients for the above experimental design.

You can either figure this question out by looking at the design matrix, or by using the contrast function from the contrast library with random numbers for y. The contrast vector will be returned as contrast(...)$X.

What should the contrast vector be, to obtain the difference between the species B control group and the species A treatment group (species B control - species A treatment)? Assume that the coefficients (columns of design matrix) are: Intercept, speciesB, condition-treated.

- A) 0 0 1

- B) 0 -1 0

- C) 0 1 1

- D) 0 1 -1

- E) 0 -1 1

- F) 1 0 1

2. Use the Rmd script to load the spider dataset. Suppose we build a model using two variables: ~ type + leg.

What is the t-statistic for the contrast of leg pair L4 vs. leg pair L2?

3. The t-statistic for the contrast of leg pair L4 vs. leg pair L2 is constructed by taking the difference of the estimated coefficients legL4 and legL2, and then dividing by the standard error of the difference.

For a contrast vector $\mathbf{C}$, the standard error of the contrast $\mathbf{C}\hat{\beta}$ is:

$$\sqrt{\mathbf{C}\hat{\mathbf{\Sigma}}\mathbf{C}^\top}$$

with $\hat{\mathbf{\Sigma}}$ the estimated covariance matrix of the coefficient estimates $\hat{\beta}$. The covariance matrix contains elements which give the variance or covariance of elements in $\hat{\beta}$. The elements on the diagonal of the $\hat{\mathbf{\Sigma}}$ matrix give the estimated variance of each element in $\hat{\beta}$. The square root of these is the standard error of the elements in $\hat{\beta}$. The off-diagonal elements of $\hat{\Sigma}$ give the estimated covariance of two different elements of $\hat{\beta}$. So $\hat{\Sigma}_{1,2}$ gives the covariance of the first and second element of $\hat{\beta}$. The $\hat{\beta}$ matrix is symmetric, which means $\hat{\Sigma}_{i,j} = \hat{\Sigma}_{j,i}$.

For the difference, we have that:

$$\mathrm{var}(\hat{\beta}_{L4} - \hat{\beta}_{L2}) = \mathrm{var}(\hat{\beta}_{L4}) + \mathrm{var}(\hat{\beta}_{L2}) - 2\mathrm{cov}(\hat{\beta}_{L4}, \hat{\beta}_{L2})$$

We showed how to estimate $\hat{\Sigma}$ using:

```
X <- model.matrix(~ type + leg, data=spider)
Sigma.hat <- sum(fitTL$residuals^2)/(nrow(X) - ncol(X)) * solve(t(X) %*% X)
```

Using the estimate of Σ, what is your estimate of $\mathrm{cov}(\hat{\beta}_{L4}, \hat{\beta}_{L2})$?
Our contrast matrix for the desired comparison is:

```
C <- matrix(c(0,0,-1,0,1),1,5)
```

4. Suppose that we notice that the within-group variances for the groups with smaller frictional coefficients are generally smaller, and so we try to apply a transformation to the frictional coefficients to make the within-group variances more constant.

Add a new variable log2friction to the spider dataframe:

```
spider$log2friction <- log2(spider$friction)
```

The Y values now look like:

```
boxplot(log2friction ~ type*leg, data=spider)
```

Run a linear model of log2friction with type, leg and interactions between type and leg.
What is the t-statistic for the interaction of type push and leg L4? If this t-statistic is sufficiently large, we would reject the null hypothesis that the push vs. pull effect on log2(friction) is the same in L4 as in L1.

5. Using the same analysis of log2 transformed data, what is the F-value for all of the type:leg interaction terms in an ANOVA? If this value is sufficiently large, we would reject the null hypothesis that the push vs. pull effect on log2(friction) is the same for all leg pairs.

6. What is the L2 vs. L1 estimate in log2(friction) for the pull samples?

7. What is the L2 vs. L1 estimate in log2(friction) for the push samples? Remember, because of the interaction terms, this is not the same as the L2 vs. L1 difference for the pull samples. If you're not sure use the contrast function. Another hint: consider the arrows plot for the model with interactions.

5.12   Collinearity

If an experiment is designed incorrectly we may not be able to estimate the parameters of interest. Similarly, when analyzing data we may incorrectly decide to use a model that can't be fit. If we are using linear models then we can detect these problems mathematically by looking for collinearity in the design matrix.

System of equations example The following system of equations:

$$a + c = 1$$
$$b - c = 1$$
$$a + b = 2$$

has more than one solution since there are an infinite number of triplets that satisfy $a = 1 - c, b = 1 + c$. Two examples are $a = 1, b = 1, c = 0$ and $a = 0, b = 2, c = 1$.

Matrix algebra approach The system of equations above can be written like this:

$$\begin{pmatrix} 1 & 0 & 1 \\ 0 & 1 & -1 \\ 1 & 1 & 0 \end{pmatrix} \begin{pmatrix} a \\ b \\ c \end{pmatrix} = \begin{pmatrix} 1 \\ 1 \\ 2 \end{pmatrix}$$

Note that the third column is a linear combination of the first two:

$$\begin{pmatrix} 1 \\ 0 \\ 1 \end{pmatrix} + -1 \begin{pmatrix} 0 \\ 1 \\ 1 \end{pmatrix} = \begin{pmatrix} 1 \\ -1 \\ 0 \end{pmatrix}$$

We say that the third column is collinear with the first 2. This implies that the system of equations can be written like this:

$$\begin{pmatrix} 1 & 0 & 1 \\ 0 & 1 & -1 \\ 1 & 1 & 0 \end{pmatrix} \begin{pmatrix} a \\ b \\ c \end{pmatrix} = a \begin{pmatrix} 1 \\ 0 \\ 1 \end{pmatrix} + b \begin{pmatrix} 0 \\ 1 \\ 1 \end{pmatrix} + c \begin{pmatrix} 1 - 0 \\ 0 - 1 \\ 1 - 1 \end{pmatrix}$$

$$= (a + c) \begin{pmatrix} 1 \\ 0 \\ 1 \end{pmatrix} + (b - c) \begin{pmatrix} 0 \\ 1 \\ 1 \end{pmatrix}$$

The third column does not add a constraint and what we really have are three equations and two unknowns: $a + c$ and $b - c$. Once we have values for those two quantities, there are an infinity number of triplets that can be used.

Collinearity and least squares Consider a design matrix $\mathbf{X}$ with two collinear columns. Here we create an extreme example in which one column is the opposite of another:

$$\mathbf{X} = \begin{pmatrix} \mathbf{1} & \mathbf{X}_1 & \mathbf{X}_2 & \mathbf{X}_3 \end{pmatrix} \text{ with, say, } \mathbf{X}_3 = -\mathbf{X}_2$$

This means that we can rewrite the residuals like this:

$$\mathbf{Y} - \{\mathbf{1}\beta_0 + \mathbf{X}_1\beta_1 + \mathbf{X}_2\beta_2 + \mathbf{X}_3\beta_3\} = \mathbf{Y} - \{\mathbf{1}\beta_0 + \mathbf{X}_1\beta_1 + \mathbf{X}_2\beta_2 - \mathbf{X}_2\beta_3\}$$
$$= \mathbf{Y} - \{\mathbf{1}\beta_0 + \mathbf{X}_1\beta_1 + \mathbf{X}_2(\beta_2 - \beta_3)\}$$

and if $\hat{\beta}_1$, $\hat{\beta}_2$, $\hat{\beta}_3$ is a least squares solution, then, for example, $\hat{\beta}_1$, $\hat{\beta}_2 + 1$, $\hat{\beta}_3 + 1$ is also a solution.

Confounding as an example Now we will demonstrate how collinearity helps us determine problems with our design using one of the most common errors made in current experimental design: confounding. To illustrate, let's use an imagined experiment in which we are interested in the effect of four treatments A, B, C and D. We assign two mice to each treatment. After starting the experiment by giving A and B to female mice, we realize there might be a sex effect. We decide to give C and D to males with hopes of estimating this effect. But can we estimate the sex effect? The described design implies the following design matrix:

$$
\begin{pmatrix}
Sex & A & B & C & D \\
0 & 1 & 0 & 0 & 0 \\
0 & 1 & 0 & 0 & 0 \\
0 & 0 & 1 & 0 & 0 \\
0 & 0 & 1 & 0 & 0 \\
1 & 0 & 0 & 1 & 0 \\
1 & 0 & 0 & 1 & 0 \\
1 & 0 & 0 & 0 & 1 \\
1 & 0 & 0 & 0 & 1
\end{pmatrix}
$$

Here we can see that sex and treatment are confounded. Specifically, the sex column can be written as a linear combination of the C and D matrices.

$$
\begin{pmatrix} Sex \\ 0 \\ 0 \\ 0 \\ 0 \\ 1 \\ 1 \\ 1 \\ 1 \end{pmatrix}
=
\begin{pmatrix} C \\ 0 \\ 0 \\ 0 \\ 0 \\ 1 \\ 1 \\ 0 \\ 0 \end{pmatrix}
+
\begin{pmatrix} D \\ 0 \\ 0 \\ 0 \\ 0 \\ 0 \\ 0 \\ 1 \\ 1 \end{pmatrix}
$$

This implies that a unique least squares estimate is not achievable.

5.13 Rank

The *rank* of a matrix columns is the number of columns that are independent of all the others. If the rank is smaller than the number of columns, then the LSE are not unique. In R, we can obtain the rank of matrix with the function `qr`, which we will describe in more detail in a following section.

```
Sex <- c(0,0,0,0,1,1,1,1)
A <-   c(1,1,0,0,0,0,0,0)
B <-   c(0,0,1,1,0,0,0,0)
C <-   c(0,0,0,0,1,1,0,0)
D <-   c(0,0,0,0,0,0,1,1)
X <- model.matrix(~Sex+A+B+C+D-1)
cat("ncol=",ncol(X),"rank=", qr(X)$rank,"\n")

## ncol= 5 rank= 4
```

Here we will not be able to estimate the effect of sex.

5.14 Removing Confounding

This particular experiment could have been designed better. Using the same number of male and female mice, we can easily design an experiment that allows us to compute the sex effect as well as all the treatment effects. Specifically, when we balance sex and treatments, the confounding is removed as demonstrated by the fact that the rank is now the same as the number of columns:

```
Sex <- c(0,1,0,1,0,1,0,1)
A <-    c(1,1,0,0,0,0,0,0)
B <-    c(0,0,1,1,0,0,0,0)
C <-    c(0,0,0,0,1,1,0,0)
D <-    c(0,0,0,0,0,0,1,1)
X <- model.matrix(~Sex+A+B+C+D-1)
cat("ncol=",ncol(X),"rank=", qr(X)$rank,"\n")

## ncol= 5 rank= 5
```

{pagebreak}

5.15 Exercises

Consider these design matrices:

$$
A = \begin{pmatrix} 1 & 0 & 0 & 0 \\ 1 & 0 & 0 & 0 \\ 1 & 1 & 1 & 0 \\ 1 & 1 & 0 & 1 \end{pmatrix} \quad
B = \begin{pmatrix} 1 & 0 & 0 & 1 \\ 1 & 0 & 1 & 1 \\ 1 & 1 & 0 & 0 \\ 1 & 1 & 1 & 0 \end{pmatrix} \quad
C = \begin{pmatrix} 1 & 0 & 0 \\ 1 & 1 & 2 \\ 1 & 2 & 4 \\ 1 & 3 & 6 \end{pmatrix}
$$

$$
D = \begin{pmatrix} 1 & 0 & 0 & 0 & 0 \\ 1 & 0 & 0 & 0 & 1 \\ 1 & 1 & 0 & 1 & 0 \\ 1 & 1 & 0 & 1 & 1 \\ 1 & 0 & 1 & 1 & 0 \\ 1 & 0 & 1 & 1 & 1 \end{pmatrix} \quad
E = \begin{pmatrix} 1 & 0 & 0 & 0 \\ 1 & 0 & 1 & 0 \\ 1 & 1 & 0 & 0 \\ 1 & 1 & 1 & 1 \end{pmatrix} \quad
F = \begin{pmatrix} 1 & 0 & 0 & 1 \\ 1 & 0 & 0 & 1 \\ 1 & 0 & 1 & 1 \\ 1 & 1 & 0 & 0 \\ 1 & 1 & 0 & 0 \\ 1 & 1 & 1 & 0 \end{pmatrix}
$$

1. Which of the above design matrices does NOT have the problem of collinearity?

2. The following exercises are advanced. Let's use the example from the lecture to visualize how there is not a single best $\hat{\beta}$, when the design matrix has collinearity of columns. An example can be made with:

```
sex <- factor(rep(c("female","male"),each=4))
trt <- factor(c("A","A","B","B","C","C","D","D"))
```

The model matrix can then be formed with:

```
X <- model.matrix( ~ sex + trt)
```

And we can see that the number of independent columns is less than the number of columns of X:

```
qr(X)$rank
```

Suppose we observe some outcome Y. For simplicity, we will use synthetic data:

```
Y <- 1:8
```

Now we will fix the value for two coefficients and optimize the remaining ones. We will fix β_{male} and β_D. Then we will find the optimal value for the remaining betas, in terms of minimizing the residual sum of squares. We find the value that minimize:

$$\sum_{i=1}^{8}\{(Y_i - X_{i,male}\beta_{male} - X_{i,D}\beta_{i,D}) - \mathbf{Z}_i\boldsymbol{\gamma})^2\}$$

where X_{male} is the male column of the design matrix, X_D is the D column, $\mathbf{Z}_i$ is a 1 by 3 matrix with the remaining column entries for unit i, and $\boldsymbol{\gamma}$ is a 3 x 1 matrix with the remaining parameters.

So all we need to do is redefine Y as $Y^* = Y - X_{male}\beta_{male} - X_D\beta_D$ and fit a linear model. The following line of code creates this variable Y^*, after fixing β_{male} to a value **a**, and β_D to a value, **b**:

```
makeYstar <- function(a,b) Y - X[,2] * a - X[,5] * b
```

Now we'll construct a function which, for a given value a and b, gives us back the sum of squared residuals after fitting the other terms.

```
fitTheRest <- function(a,b) {
  Ystar <- makeYstar(a,b)
  Xrest <- X[,-c(2,5)]
  betarest <- solve(t(Xrest) %*% Xrest) %*% t(Xrest) %*% Ystar
  residuals <- Ystar - Xrest %*% betarest
  sum(residuals^2)
}
```

What is the sum of squared residuals when the male coefficient is 1 and the D coefficient is 2, and the other coefficients are fit using the linear model solution?

3. We can apply our function `fitTheRest` to a grid of values for β_{male} and β_D, using the `outer` function in R. `outer` takes three arguments: a grid of values for the first argument, a grid of values for the second argument, and finally a function which takes two arguments.

Try it out:

```
outer(1:3,1:3,'*')
```

We can run `fitTheRest` on a grid of values, using the following code (the `Vectorize` is necessary as `outer` requires only vectorized functions):

```
outer(-2:8,-2:8,Vectorize(fitTheRest))
```

In the grid of values, what is the smallest sum of squared residuals?

5.16   The QR Factorization (Advanced)

We have seen that in order to calculate the LSE, we need to invert a matrix. In previous sections we used the function `solve`. However, solve is not a stable solution. When coding LSE computation, we use the QR decomposition.

Inverting $\mathbf{X}^\top \mathbf{X}$ Remember that to minimize the RSS:

$$(\mathbf{Y} - \mathbf{X}\beta)^\top (\mathbf{Y} - \mathbf{X}\beta)$$

We need to solve:

$$\mathbf{X}^\top \mathbf{X}\hat{\beta} = \mathbf{X}^\top \mathbf{Y}$$

The solution is:

$$\hat{\beta} = (\mathbf{X}^\top \mathbf{X})^{-1} \mathbf{X}^\top \mathbf{Y}$$

Thus, we need to compute $(\mathbf{X}^\top \mathbf{X})^{-1}$.

`solve` is numerically unstable   To demonstrate what we mean by *numerically unstable*, we construct an extreme case:

```
n <- 50;M <- 500
x <- seq(1,M,len=n)
X <- cbind(1,x,x^2,x^3)
colnames(X) <- c("Intercept","x","x2","x3")
beta <- matrix(c(1,1,1,1),4,1)
set.seed(1)
y <- X%*%beta+rnorm(n,sd=1)
```

The standard R function for inverse gives an error:

```
solve(crossprod(X)) %*% crossprod(X,y)
```

To see why this happens, look at $(\mathbf{X}^\top \mathbf{X})$

```
options(digits=4)
log10(crossprod(X))
```

```
##             Intercept      x     x2     x3
## Intercept      1.699  4.098  6.625  9.203
## x              4.098  6.625  9.203 11.810
## x2             6.625  9.203 11.810 14.434
## x3             9.203 11.810 14.434 17.070
```

Note the difference of several orders of magnitude. On a digital computer, we have a limited range of numbers. This makes some numbers seem like 0, when we also have to consider very large numbers. This in turn leads to divisions that are practically divisions by 0 errors.

The factorization The QR factorization is based on a mathematical result that tells us that we can decompose any full rank $N \times p$ matrix $\mathbf{X}$ as:

$$\mathbf{X} = \mathbf{QR}$$

with:

- $\mathbf{Q}$ a $N \times p$ matrix with $\mathbf{Q}^\top \mathbf{Q} = \mathbf{I}$
- $\mathbf{R}$ a $p \times p$ upper triangular matrix.

Upper triangular matrices are very convenient for solving system of equations.

Example of upper triangular matrix In the example below, the matrix on the left is upper triangular: it only has 0s below the diagonal. This facilitates solving the system of equations greatly:

$$\begin{pmatrix} 1 & 2 & -1 \\ 0 & 1 & 2 \\ 0 & 0 & 1 \end{pmatrix} \begin{pmatrix} a \\ b \\ c \end{pmatrix} = \begin{pmatrix} 6 \\ 4 \\ 1 \end{pmatrix}$$

We immediately know that $c = 1$, which implies that $b + 2 = 4$. This in turn implies $b = 2$ and thus $a + 4 - 1 = 6$ so $a = 3$. Writing an algorithm to do this is straight-forward for any upper triangular matrix.

Finding LSE with QR If we rewrite the equations of the LSE using $\mathbf{QR}$ instead of $\mathbf{X}$ we have:

$$\mathbf{X}^\top \mathbf{X} \boldsymbol{\beta} = \mathbf{X}^\top \mathbf{Y}$$

$$(\mathbf{QR})^\top (\mathbf{QR}) \boldsymbol{\beta} = (\mathbf{QR})^\top \mathbf{Y}$$

$$\mathbf{R}^\top (\mathbf{Q}^\top \mathbf{Q}) \mathbf{R} \boldsymbol{\beta} = \mathbf{R}^\top \mathbf{Q}^\top \mathbf{Y}$$

$$\mathbf{R}^\top \mathbf{R} \boldsymbol{\beta} = \mathbf{R}^\top \mathbf{Q}^\top \mathbf{Y}$$

$$(\mathbf{R}^\top)^{-1} \mathbf{R}^\top \mathbf{R} \boldsymbol{\beta} = (\mathbf{R}^\top)^{-1} \mathbf{R}^\top \mathbf{Q}^\top \mathbf{Y}$$

$$\mathbf{R} \boldsymbol{\beta} = \mathbf{Q}^\top \mathbf{Y}$$

$\mathbf{R}$ being upper triangular makes solving this more stable. Also, because $\mathbf{Q}^\top \mathbf{Q} = \mathbf{I}$, we know that the columns of $\mathbf{Q}$ are in the same scale which stabilizes the right side.

Now we are ready to find LSE using the QR decomposition. To solve:

$$\mathbf{R} \boldsymbol{\beta} = \mathbf{Q}^\top \mathbf{Y}$$

We use `backsolve` which takes advantage of the upper triangular nature of $\mathbf{R}$.

```
QR <- qr(X)
Q <- qr.Q( QR )
R <- qr.R( QR )
(betahat <- backsolve(R, crossprod(Q,y) ) )
```

```
##           [,1]
## [1,] 0.9038
## [2,] 1.0066
## [3,] 1.0000
## [4,] 1.0000
```

In practice, we do not need to do any of this due to the built-in `solve.qr` function:

```
QR <- qr(X)
(betahat <- solve.qr(QR, y))
```

```
##                [,1]
## Intercept 0.9038
## x         1.0066
## x2        1.0000
## x3        1.0000
```

Fitted values This factorization also simplifies the calculation for fitted values:

$$\mathbf{X}\hat{\beta} = (\mathbf{Q}\mathbf{R})\mathbf{R}^{-1}\mathbf{Q}^\top\mathbf{y} = \mathbf{Q}\mathbf{Q}^\top\mathbf{y}$$

In R, we simply do the following:

```
library(rafalib)
mypar(1,1)
plot(x,y)
fitted <- tcrossprod(Q)%*%y
lines(x,fitted,col=2)
```

Standard errors To obtain the standard errors of the LSE, we note that:

$$(\mathbf{X}^\top\mathbf{X})^{-1} = (\mathbf{R}^\top\mathbf{Q}^\top\mathbf{Q}\mathbf{R})^{-1} = (\mathbf{R}^\top\mathbf{R})^{-1}$$

The function `chol2inv` is specifically designed to find this inverse. So all we do is the following:

```
df <- length(y) - QR$rank
sigma2 <- sum((y-fitted)^2)/df
varbeta <- sigma2*chol2inv(qr.R(QR))
SE <- sqrt(diag(varbeta))
cbind(betahat,SE)
```

```
##                              SE
## Intercept 0.9038 4.508e-01
## x         1.0066 7.858e-03
## x2        1.0000 3.662e-05
## x3        1.0000 4.802e-08
```

This gives us identical results to the `lm` function.

```
summary(lm(y~0+X))$coef
```

```
##                Estimate Std. Error   t value    Pr(>|t|)
## XIntercept    0.9038 4.508e-01 2.005e+00   5.089e-02
## Xx            1.0066 7.858e-03 1.281e+02   2.171e-60
## Xx2           1.0000 3.662e-05 2.731e+04 1.745e-167
## Xx3           1.0000 4.802e-08 2.082e+07 4.559e-300
```

5.17 Going Further

Linear models can be extended in many directions. Here are some examples of extensions, which you might come across in analyzing data in the life sciences:

Robust linear models In calculating the solution and its estimated error in the standard linear model, we minimize the squared errors. This involves a sum of squares from all the data points, which means that a few *outlier* data points can have a large influence on the solution. In addition, the errors are assumed to have constant variance (called *homoskedasticity*), which might not always hold true (when this is not true, it is called *heteroskedasticity*). Therefore, methods have been developed to generate more *robust* solutions, which behave well in the presence of outliers, or when the distributional assumptions are not met. A number of these are mentioned on the robust statistics[8] page on the CRAN website. For more background, there is also a Wikipedia article[9] with references.

Generalized linear models In the standard linear model, we did not make any assumptions about the distribution of $\mathbf{Y}$, though in some cases we can gain better estimates if we know that $\mathbf{Y}$ is, for example, restricted to non-negative integers $0, 1, 2, \ldots$, or restricted to the interval $[0, 1]$. A framework for analyzing such cases is referred to as *generalized linear models*, commonly abbreviated as GLMs. The two key components of the GLM are the *link function* and a probability distribution. The link function g connects our familiar matrix product $\mathbf{X}\boldsymbol{\beta}$ to the $\mathbf{Y}$ values through:

$$\mathrm{E}(\mathbf{Y}) = g^{-1}(\mathbf{X}\boldsymbol{\beta})$$

R includes the function `glm` which fits GLMs and uses a familiar form as `lm`. Additional arguments include `family`, which can be used to specify the distributional assumption for $\mathbf{Y}$. Some examples of the use of GLMs are shown at the Quick R[10] website. There are a number of references for GLMs on the Wikipedia page[11].

Mixed effects linear models In the standard linear model, we assumed that the matrix $\mathbf{X}$ was *fixed* and not random. For example, we measured the frictional coefficients for each leg pair, and in the push and pull direction. The fact that an observation had a 1 for a given column in $\mathbf{X}$ was not random, but dictated by the experimental design. However, in the father and son heights example, we did not fix the values of the fathers' heights, but observed these (and likely these were measured with some error). A framework for studying the effect of the randomness for various columns in X is referred to as *mixed effects* models, which implies that some effects are *fixed* and some effects are *random*. One of the most popular packages in R for fitting linear mixed effects models is lme4[12] which has an accompanying paper on Fitting Linear Mixed-Effects Models using lme4[13]. There is also a Wikipedia page[14] with more references.

Bayesian linear models The approach presented here assumed $\boldsymbol{\beta}$ was a fixed (non-random) parameter. We presented methodology that estimates this parameter, along with

---

[8]http://cran.r-project.org/web/views/Robust.html
[9]http://en.wikipedia.org/wiki/Robust_regression
[10]http://www.statmethods.net/advstats/glm.html
[11]http://en.wikipedia.org/wiki/Generalized_linear_model
[12]http://lme4.r-forge.r-project.org/
[13]http://arxiv.org/abs/1406.5823
[14]http://en.wikipedia.org/wiki/Mixed_model

standard errors that quantify uncertainty, in the estimation process. This is referred to as the *frequentist* approach. An alternative approach is to assume that $\boldsymbol{\beta}$ is random and its distribution quantifies our prior beliefs about what $\boldsymbol{\beta}$ should be. Once we have observed data, then we update our prior beliefs by computing the conditional distribution, referred to as the *posterior* distribution, of $\boldsymbol{\beta}$ given the data. This is referred to as the *Bayesian* approach. For example, once we have computed the posterior distribution of $\boldsymbol{\beta}$ we can report the most likely outcome of an interval that occurs with high probability (credible interval). In addition, many models can be connected together in what is referred to as a *hierarchical model*. Note that we provide a brief introduction to Bayesian statistics and hierarchical models in a later chapter. A good reference for Bayesian hierarchical models is Bayesian Data Analysis[15], and some software for computing Bayesian linear models can be found on the Bayes[16] page on CRAN. Some well known software for computing Bayesian models are stan[17] and BUGS[18].

Penalized linear models Note that if we include enough parameters in a model we can achieve a residual sum of squares of 0. Penalized linear models introduce a penalty term to the least square equation we minimize. These penalities are typically of the form, $\lambda \sum_{j=1}^{p} \|\beta_j\|^k$ and they penalize for large absolute values of β as well as large numbers of parameters. The motivation for this extra term is to avoid over-fitting. To use these models, we need to pick λ which determines how much we penalize. When $k = 2$, this is referred to as *ridge* regression, Tikhonov regularization, or L2 regularization. When $k = 1$, this is referred to as *LASSO* or L1 regularization. A good reference for these penalized linear models is the Elements of Statistical Learning[19] textbook, which is available as a free pdf. Some R packages which implement penalized linear models are the lm.ridge[20] function in the MASS package, the lars[21] package, and the glmnet[22] package.

---

[15]http://www.stat.columbia.edu/~gelman/book/

[16]http://cran.r-project.org/web/views/Bayesian.html

[17]http://mc-stan.org/

[18]http://www.mrc-bsu.cam.ac.uk/software/bugs/

[19]http://statweb.stanford.edu/~tibs/ElemStatLearn/

[20]https://stat.ethz.ch/R-manual/R-devel/library/MASS/html/lm.ridge.html

[21]http://cran.r-project.org/web/packages/lars/index.html

[22]http://cran.r-project.org/web/packages/glmnet/index.html

6

Inference for High Dimensional Data

6.1 Introduction

High-throughput technologies have changed basic biology and the biomedical sciences from data poor disciplines to data intensive ones. A specific example comes from research fields interested in understanding gene expression. Gene expression is the process in which DNA, the blueprint for life, is copied into RNA, the templates for the synthesis of proteins, the building blocks for life. In the 1990s, the analysis of gene expression data amounted to spotting black dots on a piece of paper or extracting a few numbers from standard curves. With high-throughput technologies, such as microarrays, this suddenly changed to sifting through tens of thousands of numbers. More recently, RNA sequencing has further increased data complexity. Biologists went from using their eyes or simple summaries to categorize results, to having thousands (and now millions) of measurements per sample to analyze. In this chapter, we will focus on statistical inference in the context of high-throughput measurements. Specifically, we focus on the problem of detecting differences in groups using statistical tests and quantifying uncertainty in a meaningful way. We also introduce exploratory data analysis techniques that should be used in conjunction with inference when analyzing high-throughput data. In later chapters, we will study the statistics behind clustering, machine learning, factor analysis and multi-level modeling.

Since there is a vast number of available public datasets, we use several gene expression examples. Nonetheless, the statistical techniques you will learn have also proven useful in other fields that make use of high-throughput technologies. Technologies such as microarrays, next generation sequencing, fRMI, and mass spectrometry all produce data to answer questions for which what we learn here will be indispensable.

Data packages Several of the examples we are going to use in the following sections are best obtained through R packages. These are available from GitHub and can be installed using the `install_github` function from the `devtools` package. Microsoft Windows users might need to follow these instructions[1] to properly install `devtools`.

Once `devtools` is installed, you can then install the data packages like this:

```
library(devtools)
install_github("genomicsclass/GSE5859Subset")
```

The three tables Most of the data we use as examples in this book are created with high-throughput technologies. These technologies measure thousands of *features*. Examples of features are genes, single base locations of the genome, genomic regions, or image pixel intensities. Each specific measurement product is defined by a specific set of features. For example, a specific gene expression microarray product is defined by the set of genes that it measures.

---

[1]https://github.com/genomicsclass/windows

A specific study will typically use one product to make measurements on several experimental units, such as individuals. The most common experimental unit will be the individual, but they can also be defined by other entities, for example different parts of a tumor. We often call the experimental units *samples* following experimental jargon. It is important that these are not confused with samples as referred to in previous chapters, for example "random sample".

So a high-throughput experiment is usually defined by three tables: one with the high-throughput measurements and two tables with information about the columns and rows of this first table respectively.

Because a dataset is typically defined by a set of experimental units and a product defines a fixed set of features, the high-throughput measurements can be stored in an $n \times m$ matrix, with n the number of units and m the number of features. In R, the convention has been to store the transpose of these matrices.

Here is an example from a gene expression dataset:

```
library(GSE5859Subset)
data(GSE5859Subset) ##this loads the three tables
dim(geneExpression)
```

```
## [1] 8793    24
```

We have RNA expression measurements for 8793 genes from blood taken from 24 individuals (the experimental units). For most statistical analyses, we will also need information about the individuals. For example, in this case the data was originally collected to compare gene expression across ethnic groups. However, we have created a subset of this dataset for illustration and separated the data into two groups:

```
dim(sampleInfo)
```

```
## [1] 24   4
```

```
head(sampleInfo)
```

```
##     ethnicity       date         filename group
## 107       ASN 2005-06-23 GSM136508.CEL.gz     1
## 122       ASN 2005-06-27 GSM136530.CEL.gz     1
## 113       ASN 2005-06-27 GSM136517.CEL.gz     1
## 163       ASN 2005-10-28 GSM136576.CEL.gz     1
## 153       ASN 2005-10-07 GSM136566.CEL.gz     1
## 161       ASN 2005-10-07 GSM136574.CEL.gz     1
```

```
sampleInfo$group
```

```
##  [1] 1 1 1 1 1 1 1 1 1 1 1 1 0 0 0 0 0 0 0 0 0 0 0 0
```

One of the columns, filenames, permits us to connect the rows of this table to the columns of the measurement table.

```
match(sampleInfo$filename,colnames(geneExpression))
```

```
##  [1]  1  2  3  4  5  6  7  8  9 10 11 12 13 14 15 16 17 18 19 20 21 22 23
## [24] 24
```

Finally, we have a table describing the features:

```
dim(geneAnnotation)
```

```
## [1] 8793    4
```

```
head(geneAnnotation)
```

```
##      PROBEID  CHR    CHRLOC SYMBOL
## 1   1007_s_at chr6  30852327   DDR1
## 30   1053_at chr7 -73645832   RFC2
## 31    117_at chr1 161494036  HSPA6
## 32    121_at chr2 -113973574  PAX8
## 33 1255_g_at chr6  42123144 GUCA1A
## 34   1294_at chr3 -49842638   UBA7
```

The table includes an ID that permits us to connect the rows of this table with the rows of the measurement table:

```
head(match(geneAnnotation$PROBEID,rownames(geneExpression)))
```

```
## [1] 1 2 3 4 5 6
```

The table also includes biological information about the features, namely chromosome location and the gene "name" used by biologists.

6.2 Exercises

For the remaining parts of this book we will be downloading larger datasets than those we have been using. Most of these datasets are not available as part of the standard R installation or packages such as `UsingR`. For some of these packages, we have created packages and offer them via GitHub. To download these you will need to install the `devtools` package. Once you do this, you can install packages such as the `GSE5859Subset` which we will be using here:

```
library(devtools)
install_github("genomicsclass/GSE5859Subset")
library(GSE5859Subset)
data(GSE5859Subset)
```

This package loads three tables: `geneAnnotation`, `geneExpression`, and `sampleInfo`. Answer the following questions to familiarize yourself with the data set:

1. How many samples where processed on 2005-06-27?

2. Question: How many of the genes represented in this particular technology are on chromosome Y?

3. What is the log expression value of the gene ARPC1A on the one subject that we measured on 2005-06-10?

6.3   Inference in Practice

Suppose we were given high-throughput gene expression data that was measured for several individuals in two populations. We are asked to report which genes have different average expression levels in the two populations. If instead of thousands of genes, we were handed data from just one gene, we could simply apply the inference techniques that we have learned before. We could, for example, use a t-test or some other test. Here we review what changes when we consider high-throughput data.

p-values are random variables An important concept to remember in order to understand the concepts presented in this chapter is that p-values are random variables. To see this, consider the example in which we define a p-value from a t-test with a large enough sample size to use the CLT approximation. Then our p-value is defined as the probability that a normally distributed random variable is larger, in absolute value, than the observed t-test, call it Z. So for a two sided test the p-value is:

$$p = 2\{1 - \Phi(|Z|)\}$$

In R, we write:

```
2*( 1-pnorm( abs(Z) ) )
```

Now because Z is a random variable and Φ is a deterministic function, p is also a random variable. We will create a Monte Carlo simulation showing how the values of p change. We use `femaleControlsPopulation.csv` from earlier chapters.

We read in the data, and use `replicate` to repeatedly create p-values.

```
set.seed(1)
population = unlist( read.csv(filename) )
N <- 12
B <- 10000
pvals <- replicate(B,{
  control = sample(population,N)
  treatment = sample(population,N)
  t.test(treatment,control)$p.val
  })
hist(pvals)
```

As implied by the histogram, in this case the distribution of the p-value is uniformly distributed. In fact, we can show theoretically that when the null hypothesis is true, this is always the case. For the case in which we use the CLT, we have that the null hypothesis H_0 implies that our test statistic Z follows a normal distribution with mean 0 and SD 1 thus:

$$p_a = \Pr(Z < a \mid H_0) = \Phi(a)$$

This implies that:

$$\Pr(p < p_a) = \Pr[\Phi^{-1}(p) < \Phi^{-1}(p_a)]$$
$$= \Pr(Z < a) = p_a$$

which is the definition of a uniform distribution.

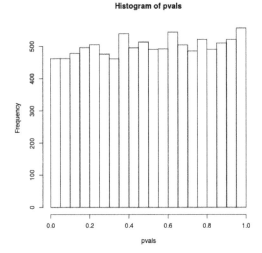

FIGURE 6.1
P-value histogram for 10,000 tests in which null hypothesis is true.

Thousands of tests In this data we have two groups denoted with 0 and 1:

```
library(GSE5859Subset)
data(GSE5859Subset)
g <- sampleInfo$group
g
```

```
## [1] 1 1 1 1 1 1 1 1 1 1 1 1 0 0 0 0 0 0 0 0 0 0 0 0
```

If we were interested in a particular gene, let's arbitrarily pick the one on the 25th row, we would simply compute a t-test. To compute a p-value, we will use the t-distribution approximation and therefore we need the population data to be approximately normal. We check this assumption with a qq-plot:

```
e <- geneExpression[25,]
```

```
library(rafalib)
mypar(1,2)
```

```
qqnorm(e[g==1])
qqline(e[g==1])
```

```
qqnorm(e[g==0])
qqline(e[g==0])
```

The qq-plots show that the data is well approximated by the normal approximation. The t-test does not find this gene to be statistically significant:

```
t.test(e[g==1],e[g==0])$p.value
```

```
## [1] 0.779303
```

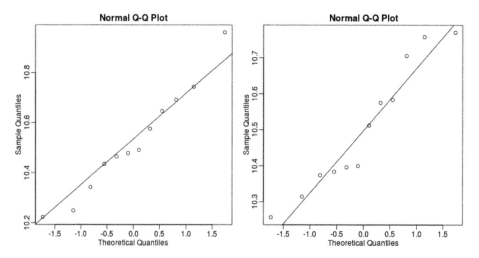

FIGURE 6.2
Normal qq-plots for one gene. Left plot shows first group and right plot shows second group.

To answer the question for each gene, we simply repeat the above for each gene. Here we will define our own function and use `apply`:

```
myttest <- function(x) t.test(x[g==1],x[g==0],var.equal=TRUE)$p.value
pvals <- apply(geneExpression,1,myttest)
```

We can now see which genes have p-values less than, say, 0.05. For example, right away we see that...

```
sum(pvals<0.05)
```

```
## [1] 1383
```

... genes had p-values less than 0.05.

However, as we will describe in more detail below, we have to be careful in interpreting this result because we have performed over 8,000 tests. If we performed the same procedure on random data, for which the null hypothesis is true for all features, we obtain the following results:

```
set.seed(1)
m <- nrow(geneExpression)
n <- ncol(geneExpression)
randomData <- matrix(rnorm(n*m),m,n)
nullpvals <- apply(randomData,1,myttest)
sum(nullpvals<0.05)
```

```
## [1] 419
```

As we will explain later in the chapter, this is to be expected: 419 is roughly 0.05*8192 and we will describe the theory that tells us why this prediction works.

Faster t-test implementation Before we continue, we should point out that the above implementation is very inefficient. There are several faster implementations that perform t-test for high-throughput data. We make use of a function that is not available from CRAN, but rather from the Bioconductor project.

To download and install packages from Bioconductor, we can use the `install_bioc` function in **rafalib** to install the package:

```
install_bioc("genefilter")
```

Now we can show that this function is much faster than our code above and produce practically the same answer:

```
library(genefilter)
results <- rowttests(geneExpression,factor(g))
max(abs(pvals-results$p))
```

```
## [1] 6.528111e-14
```

6.4   Exercises

These exercises will help clarify that p-values are random variables and some of the properties of these p-values. Remember that, as the sample average is a random variable because it is based on a random sample, the p-values are also random variables as they are based themselves on random variables: the sample mean and sample standard deviation for example.

To see this, let's see how p-values change when we take different samples.

```
set.seed(1)
dir = "https://raw.githubusercontent.com/genomicsclass/dagdata/master/inst/extdata/"
filename = "femaleControlsPopulation.csv"
url = paste0(dir, filename)
population = read.csv(url)
pvals <- replicate(1000,{
  control = sample(population[,1],12)
  treatment = sample(population[,1],12)
  t.test(treatment,control)$p.val
})
head(pvals)
hist(pvals)
```

1. What proportion of the p-values is below 0.05?

2. What proportion of the p-values is below 0.01?

3. Assume you are testing the effectiveness of 20 diets on mice weight. For each of the 20 diets, you run an experiment with 10 control mice and 10 treated mice. Assume the null hypothesis, that the diet has no effect, is true for all 20 diets and that mice weights follow a normal distribution, with mean 30 grams and a standard deviation of 2 grams. Run a Monte Carlo simulation for one of these studies:

```
cases = rnorm(10,30,2)
controls = rnorm(10,30,2)
t.test(cases,controls)
```

Question: Now run a Monte Carlo simulation imitating the results for the experiment for all 20 diets. If you set the seed at 100, `set.seed(100)`, how many of p-values are below 0.05?

4. Now create a simulation to learn about the distribution of the number of p-values that are less than 0.05. In question 3, we ran the 20 diet experiment once. Now we will run the experiment 1,000 times and each time save the number of p-values that are less than 0.05.

Set the seed at 100, `set.seed(100)`, run the code from Question 3 1,000 times, and save the number of times the p-value is less than 0.05 for each of the 1,000 instances. What is the average of these numbers? This is the expected number of tests (out of the 20 we run) that we will reject when the null is true.

5. What this says is that on average, we expect some p-value to be 0.05 even when the null is true for all diets. Use the same simulation data and report for what percent of the 1,000 replications did we make more than 0 false positives?

6.5 Procedures

In the previous section we learned how p-values are no longer a useful quantity to interpret when dealing with high-dimensional data. This is because we are testing many *features* at the same time. We refer to this as the *multiple comparison* or *multiple testing* or *multiplicity* problem. The definition of a p-value does not provide a useful quantification here. Again, because when we test many hypotheses simultaneously, a list based simply on a small p-value cut-off of, say 0.01, can result in many false positives with high probability. Here we define terms that are more appropriate in the context of high-throughput data.

The most widely used approach to the multiplicity problem is to define a *procedure* and then estimate or *control* an informative *error rate* for this procedure. What we mean by *control* here is that we adapt the procedure to guarantee an *error rate* below a predefined value. The procedures are typically flexible through parameters or cutoffs that let us control specificity and sensitivity. An example of a procedure is:

- Compute a p-value for each gene.
- Call significant all genes with p-values smaller than α.

Note that changing the α permits us to adjust specificity and sensitivity.
Next we define the *error rates* that we will try to estimate and control.

6.6 Error Rates

Throughout this section we will be using the type I error and type II error terminology. We will also refer to them as false positives and false negatives respectively. We also use the more general terms specificity, which relates to type I error, and sensitivity, which relates to type II errors.

In the context of high-throughput data we can make several type I errors and several type II errors in one experiment, as opposed to one or the other as seen in Chapter 1. In this table, we summarize the possibilities using the notation from the seminal paper by Benjamini-Hochberg:

	Called significant	Not called significant	Total
Null True	V	$m_0 - V$	m_0
Alternative True	S	$m_1 - S$	m_1
True	R	$m - R$	m

To describe the entries in the table, let's use as an example a dataset representing measurements from 10,000 genes, which means that the total number of tests that we are conducting is: $m = 10,000$. The number of genes for which the null hypothesis is true, which in most cases represent the "non-interesting" genes, is m_0, while the number of genes for which the null hypothesis is false is m_1. For this we can also say that the *alternative hypothesis* is true. In general, we are interested in *detecting* as many as possible of the cases for which the alternative hypothesis is true (true positives), without incorrectly detecting cases for which the null hypothesis is true (false positives). For most high-throughput experiments, we assume that m_0 is much greater than m_1. For example, we test 10,000 expecting 100 genes or less to be *interesting*. This would imply that $m_1 \leq 100$ and $m_0 \geq 19,900$.

Throughout this chapter we refer to *features* as the units being tested. In genomics, examples of features are genes, transcripts, binding sites, CpG sites, and SNPs. In the table, R represents the total number of features that we call significant after applying our procedure, while $m - R$ is the total number of genes we don't call significant. The rest of the table contains important quantities that are unknown in practice.

- V represents the number of type I errors or false positives. Specifically, V is the number of features for which the null hypothesis is true, that we call significant.
- S represents the number of true positives. Specifically, S is the number of features for which the alternative is true, that we call significant.

This implies that there are $m_1 - S$ type II errors or *false negatives* and $m_0 - V$ true negatives. Keep in mind that if we only ran one test, a p-value is simply the probability that $V = 1$ when $m = m_0 = 1$. Power is the probability of $S = 1$ when $m = m_1 = 1$. In this very simple case, we wouldn't bother making the table above, but now we show how defining the terms in the table helps for the high-dimensional setting.

Data example Let's compute these quantities with a data example. We will use a Monte Carlo simulation using our mice data to imitate a situation in which we perform tests for 10,000 different fad diets, none of them having an effect on weight. This implies that the null hypothesis is true for diets and thus $m = m_0 = 10,000$ and $m_1 = 0$. Let's run the tests with a sample size of $N = 12$ and compute R. Our procedure will declare any diet achieving a p-value smaller than $\alpha = 0.05$ as significant.

```
set.seed(1)
population = unlist( read.csv("femaleControlsPopulation.csv") )
alpha <- 0.05
N <- 12
m <- 10000
pvals <- replicate(m,{
  control = sample(population,N)
  treatment = sample(population,N)
```

```
  t.test(treatment,control)$p.value
})
```

Although in practice we do not know the fact that no diet works, in this simulation we do, and therefore we can actually compute V and S. Because all null hypotheses are true, we know, in this specific simulation, that $V = R$. Of course, in practice we can compute R but not V.

```
sum(pvals < 0.05) ##This is R
```

```
## [1] 462
```

These many false positives are not acceptable in most contexts.

Here is more complicated code showing results where 10% of the diets are effective with an average effect size of $\Delta = 3$ ounces. Studying this code carefully will help us understand the meaning of the table above. First let's define *the truth*:

```
alpha <- 0.05
N <- 12
m <- 10000
p0 <- 0.90 ##10% of diets work, 90% don't
m0 <- m*p0
m1 <- m-m0
nullHypothesis <- c( rep(TRUE,m0), rep(FALSE,m1))
delta <- 3
```

Now we are ready to simulate 10,000 tests, perform a t-test on each, and record if we rejected the null hypothesis or not:

```
set.seed(1)
calls <- sapply(1:m, function(i){
  control <- sample(population,N)
  treatment <- sample(population,N)
  if(!nullHypothesis[i]) treatment <- treatment + delta
  ifelse( t.test(treatment,control)$p.value < alpha,
          "Called Significant",
          "Not Called Significant")
})
```

Because in this simulation we know the truth (saved in `nullHypothesis`), we can compute the entries of the table:

```
null_hypothesis <- factor( nullHypothesis, levels=c("TRUE","FALSE"))
table(null_hypothesis,calls)
```

```
##                   calls
## null_hypothesis Called Significant Not Called Significant
##           TRUE                 421                   8579
##           FALSE                520                    480
```

The first column of the table above shows us V and S. Note that V and S are random variables. If we run the simulation repeatedly, these values change. Here is a quick example:

```
B <- 10 ##number of simulations
VandS <- replicate(B,{
  calls <- sapply(1:m, function(i){
    control <- sample(population,N)
    treatment <- sample(population,N)
    if(!nullHypothesis[i]) treatment <- treatment + delta
    t.test(treatment,control)$p.val < alpha
  })
  cat("V =",sum(nullHypothesis & calls), "S =",sum(!nullHypothesis & calls),"\n")
  c(sum(nullHypothesis & calls),sum(!nullHypothesis & calls))
  })
## V = 410 S = 564
## V = 400 S = 552
## V = 366 S = 546
## V = 382 S = 553
## V = 372 S = 505
## V = 382 S = 530
## V = 381 S = 539
## V = 396 S = 554
## V = 380 S = 550
## V = 405 S = 569
```

This motivates the definition of error rates. We can, for example, estimate probability that V is larger than 0. This is interpreted as the probability of making at least one type I error among the 10,000 tests. In the simulation above, V was much larger than 1 in every single simulation, so we suspect this probability is very practically 1. When $m = 1$, this probability is equivalent to the p-value. When we have a multiple tests situation, we call it the Family Wide Error Rate (FWER) and it relates to a technique that is widely used: The Bonferroni Correction.

6.7   The Bonferroni Correction

Now that we have learned about the Family Wide Error Rate (FWER), we describe what we can actually do to control it. In practice, we want to choose a *procedure* that guarantees the FWER is smaller than a predetermined value such as 0.05. We can keep it general and instead of 0.05, use α in our derivations.

Since we are now describing what we do in practice, we no longer have the advantage of knowing *the truth*. Instead, we pose a procedure and try to estimate the FWER. Let's consider the naive procedure: "reject all the hypotheses with p-value <0.01". For illustrative purposes we will assume all the tests are independent (in the case of testing diets this is a safe assumption; in the case of genes it is not so safe since some groups of genes act together). Let $p_1, \ldots, p_{10000}$ be the the p-values we get from each test. These are independent random variables so:

$$\Pr(\text{at least one rejection}) = 1 - \Pr(\text{no rejections})$$

$$= 1 - \prod_{i=1}^{1000} \Pr(p_i > 0.01)$$

$$= 1 - 0.95^{1000} \approx 1$$

Or if you want to use simulations:

```
B<-10000
minpval <- replicate(B, min(runif(10000,0,1))<0.01)
mean(minpval>=1)
```

```
## [1] 1
```

So our FWER is 1! This is not what we were hoping for. If we wanted it to be lower than $\alpha = 0.05$, we failed miserably.

So what do we do to make the probability of a mistake lower than α? Using the derivation above we can change the procedure by selecting a more stringent cutoff, previously 0.01, to lower our probability of at least one mistake to be 5%. Namely, by noting that:

$$\Pr(\text{at least one rejection}) = 1 - (1 - k)^{10000}$$

and solving for k, we get $1 - (1 - k)^{10000} = 0.05 \implies k = 1 - 0.95^{1/10000} \approx 0.000005$

This now gives a specific example of a *procedure*. This one is actually called Sidak's procedure. Specifically, we define a set of instructions, such as "reject all the null hypothesis for which p-values < 0.000005". Then, knowing the p-values are random variables, we use statistical theory to compute how many mistakes, on average, we are expected to make if we follow this procedure. More precisely, we compute bounds on these rates; that is, we show that they are smaller than some predetermined value. There is a preference in the life sciences to err on the side of being conservative.

A problem with Sidak's procedure is that it assumes the tests are independent. It therefore only controls FWER when this assumption holds. The Bonferroni correction is more general in that it controls FWER even if the tests are not independent. As with Sidak's procedure we start by noting that:

$$FWER = \Pr(V > 0) \leq \Pr(V > 0 \mid \text{all nulls are true})$$

or using the notation from the table above:

$$\Pr(V > 0) \leq \Pr(V > 0 \mid m_1 = 0)$$

The Bonferroni procedure sets $k = \alpha/m$ since we can show that:

$$\Pr(V > 0 \mid m_1 = 0) = \Pr\left(\min_i\{p_i\} \leq \frac{\alpha}{m} \mid m_1 = 0\right)$$

$$\leq \sum_{i=1}^{m} \Pr\left(p_i \leq \frac{\alpha}{m}\right)$$

$$= m\frac{\alpha}{m} = \alpha$$

Controlling the FWER at 0.05 is a very conservative approach. Using the p-values computed in the previous section...

```
set.seed(1)
pvals <- sapply(1:m, function(i){
  control <- sample(population,N)
  treatment <- sample(population,N)
  if(!nullHypothesis[i]) treatment <- treatment + delta
  t.test(treatment,control)$p.value
})
```

...we note that only:

```
sum(pvals < 0.05/10000)
```

```
## [1] 2
```

are called significant after applying the Bonferroni procedure, despite having 1,000 diets that work.

6.8   False Discovery Rate

There are many situations for which requiring an FWER of 0.05 does not make sense as it is much too strict. For example, consider the very common exercise of running a preliminary small study to determine a handful of candidate genes. This is referred to as a *discovery* driven project or experiment. We may be in search of an unknown causative gene and more than willing to perform follow-up studies with many more samples on just the candidates. If we develop a procedure that produces, for example, a list of 10 genes of which 1 or 2 pan out as important, the experiment is a resounding success. With a small sample size, the only way to achieve a FWER ≤ 0.05 is with an empty list of genes. We already saw in the previous section that despite 1,000 diets being effective, we ended up with a list with just 2. Change the sample size to 6 and you very likely get 0:

```
set.seed(1)
pvals <- sapply(1:m, function(i){
  control <- sample(population,6)
  treatment <- sample(population,6)
  if(!nullHypothesis[i]) treatment <- treatment + delta
  t.test(treatment,control)$p.value
  })
sum(pvals < 0.05/10000)
```

```
## [1] 0
```

By requiring a FWER ≤ 0.05, we are practically assuring 0 power (sensitivity). In many applications, this specificity requirement is over-kill. A widely used alternative to the FWER is the false discover rate (FDR). The idea behind FDR is to focus on the random variable $Q \equiv V/R$ with $Q = 0$ when $R = 0$ and $V = 0$. Note that $R = 0$ (nothing called significant) implies $V = 0$ (no false positives). So Q is a random variable that can take values between 0 and 1 and we can define a rate by considering the average of Q. To better understand this concept here, we compute Q for the procedure: call everything p-value < 0.05 significant.

Vectorizing code Before running the simulation, we are going to *vectortize* the code. This means that instead of using `sapply` to run `m` tests, we will create a matrix with all data in one call to sample. This code runs several times faster than the code above, which is necessary here due to the fact that we will be generating several simulations. Understanding this chunk of code and how it is equivalent to the code above using `sapply` will take a you long way in helping you code efficiently in R.

```
library(genefilter) ##rowttests is here
set.seed(1)
```

```
##Define groups to be used with rowttests
g <- factor( c(rep(0,N),rep(1,N)) )
B <- 1000 ##number of simulations
Qs <- replicate(B,{
  ##matrix with control data (rows are tests, columns are mice)
  controls <- matrix(sample(population, N*m, replace=TRUE),nrow=m)

  ##matrix with control data (rows are tests, columns are mice)
  treatments <-  matrix(sample(population, N*m, replace=TRUE),nrow=m)

  ##add effect to 10% of them
  treatments[which(!nullHypothesis),]<-treatments[which(!nullHypothesis),]+delta

  ##combine to form one matrix
  dat <- cbind(controls,treatments)

 calls <- rowttests(dat,g)$p.value < alpha
 R=sum(calls)
 Q=ifelse(R>0,sum(nullHypothesis & calls)/R,0)
 return(Q)
})
```

Controlling FDR The code above is a Monte Carlo simulation that generates 10,000 experiments 1,000 times, each time saving the observed Q. Here is a histogram of these values:

```
library(rafalib)
mypar(1,1)
hist(Qs) ##Q is a random variable, this is its distribution
```

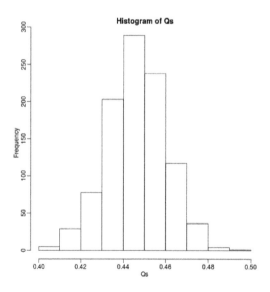

FIGURE 6.3
Q (false positives divided by number of features called significant) is a random variable. Here we generated a distribution with a Monte Carlo simulation.

The FDR is the average value of Q

```
FDR=mean(Qs)
print(FDR)
```

```
## [1] 0.4463354
```

The FDR is relatively high here. This is because for 90% of the tests, the null hypotheses is true. This implies that with a 0.05 p-value cut-off, out of the 100 tests we incorrectly call between 4 and 5 significant on average. This combined with the fact that we don't "catch" all the cases where the alternative is true, gives us a relatively high FDR. So how can we control this? What if we want lower FDR, say 5%?

To visually see why the FDR is high, we can make a histogram of the p-values. We use a higher value of m to have more data from the histogram. We draw a horizontal line representing the uniform distribution one gets for the m0 cases for which the null is true.

```
set.seed(1)
controls <- matrix(sample(population, N*m, replace=TRUE),nrow=m)
treatments <-  matrix(sample(population, N*m, replace=TRUE),nrow=m)
treatments[which(!nullHypothesis),]<-treatments[which(!nullHypothesis),]+delta
dat <- cbind(controls,treatments)
pvals <- rowttests(dat,g)$p.value

h <- hist(pvals,breaks=seq(0,1,0.05))
polygon(c(0,0.05,0.05,0),c(0,0,h$counts[1],h$counts[1]),col="grey")
abline(h=m0/20)
```

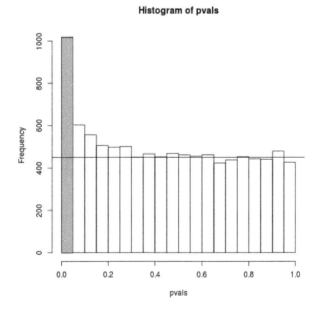

FIGURE 6.4
Histogram of p-values. Monte Carlo simulation was used to generate data with m_1 genes having differences between groups.

The first bar (grey) on the left represents cases with p-values smaller than 0.05. From the horizontal line we can infer that about 1/2 are false positives. This is in agreement with an FDR of 0.50. If we look at the bar for 0.01, we can see a lower FDR, as expected, but would call fewer features significant.

```
h <- hist(pvals,breaks=seq(0,1,0.01))
polygon(c(0,0.01,0.01,0),c(0,0,h$counts[1],h$counts[1]),col="grey")
abline(h=m0/100)
```

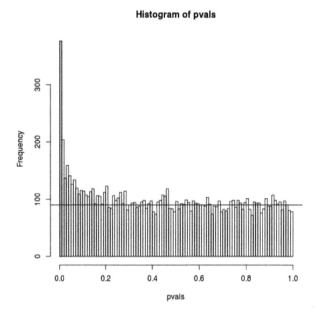

FIGURE 6.5
Histogram of p-values with breaks at every 0.01. Monte Carlo simulation was used to generate data with m_1 genes having differences between groups.

As we consider a lower and lower p-value cut-off, the number of features detected decreases (loss of sensitivity), but our FDR also decreases (gain of specificity). So how do we decide on this cut-off? One approach is to set a desired FDR level α, and then develop procedures that control the error rate: FDR $\leq \alpha$.

Benjamini-Hochberg (Advanced) We want to construct a procedure that guarantees the FDR to be below a certain level α. For any given α, the Benjamini-Hochberg (1995) procedure is very practical because it simply requires that we are able to compute p-values for each of the individual tests and this permits a procedure to be defined.

For this procedure, order the p-values in increasing order: $p_{(1)}, \ldots, p_{(m)}$. Then define k to be the largest i for which

$$p_{(i)} \leq \frac{i}{m}\alpha$$

The procedure is to reject tests with p-values smaller or equal to $p_{(k)}$. Here is an example of how we would select the k with code using the p-values computed above:

```
alpha <- 0.05
i = seq(along=pvals)
```

```
mypar(1,2)
plot(i,sort(pvals))
abline(0,i/m*alpha)
##close-up
plot(i[1:15],sort(pvals)[1:15],main="Close-up")
abline(0,i/m*alpha)
```

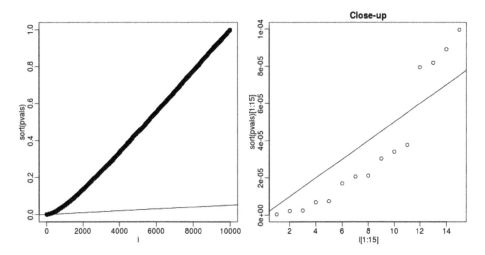

FIGURE 6.6
Plotting p-values plotted against their rank illustrates the Benjamini-Hochberg procedure. The plot on the right is a close-up of the plot on the left.

```
k <- max( which( sort(pvals) < i/m*alpha) )
cutoff <- sort(pvals)[k]
cat("k =",k,"p-value cutoff=",cutoff)
```

```
## k = 11 p-value cutoff= 3.763357e-05
```

We can show mathematically that this procedure has FDR lower than 5%. Please see Benjamini-Hochberg (1995) for details. An important outcome is that we now have selected 11 tests instead of just 2. If we are willing to set an FDR of 50% (this means we expect at least 1/2 our genes to be hits), then this list grows to 1063. The FWER does not provide this flexibility since any list of substantial size will result in an FWER of 1.

Keep in mind that we don't have to run the complicated code above as we have functions to do this. For example, using the p-values `pvals` computed above, we simply type the following:

```
fdr <- p.adjust(pvals, method="fdr")
mypar(1,1)
plot(pvals,fdr,log="xy")
abline(h=alpha,v=cutoff) ##cutoff was computed above
```

We can run a Monte-Carlo simulation to confirm that the FDR is in fact lower than .05. We compute all p-values first, and then use these to decide which get called.

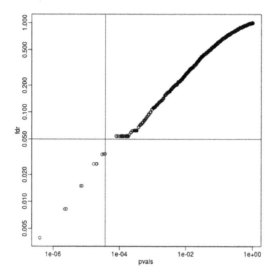

FIGURE 6.7
FDR estimates plotted against p-value.

```
alpha <- 0.05
B <- 1000 ##number of simulations. We should increase for more precision
res <- replicate(B,{
  controls <- matrix(sample(population, N*m, replace=TRUE),nrow=m)
  treatments <-  matrix(sample(population, N*m, replace=TRUE),nrow=m)
  treatments[which(!nullHypothesis),]<-treatments[which(!nullHypothesis),]+delta
  dat <- cbind(controls,treatments)
  pvals <- rowttests(dat,g)$p.value
  ##then the FDR
  calls <- p.adjust(pvals,method="fdr") < alpha
  R=sum(calls)
  Q=ifelse(R>0,sum(nullHypothesis & calls)/R,0)
  return(c(R,Q))
})
Qs <- res[2,]
mypar(1,1)
hist(Qs) ##Q is a random variable, this is its distribution


FDR=mean(Qs)
print(FDR)


## [1] 0.03813818
```

The FDR is lower than 0.05. This is to be expected because we need to be conservative to ensure the FDR ≤ 0.05 for any value of m_0, such as for the extreme case where every hypothesis tested is null: $m = m_0$. If you re-do the simulation above for this case, you will find that the FDR increases.

We should also note that in ...

```
Rs <- res[1,]
mean(Rs==0)*100
```

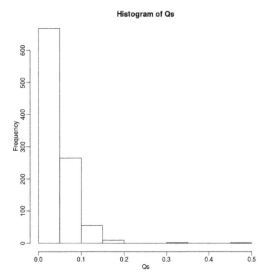

FIGURE 6.8
Histogram of Q (false positives divided by number of features called significant) when the alternative hypothesis is true for some features.

```
## [1] 0.7
```

... percent of the simulations, we did not call any genes significant.

Finally, note that the `p.adjust` function has several options for error rate controlling procedures:

```
p.adjust.methods
```

```
## [1] "holm"       "hochberg"  "hommel"      "bonferroni" "BH"
## [6] "BY"         "fdr"       "none"
```

It is important to remember that these options offer not just different approaches to estimating error rates, but also that different error rates are estimated: namely FWER and FDR. This is an important distinction. More information is available from:

```
?p.adjust
```

In summary, requiring that FDR $\leq$ 0.05 is a much more lenient requirement FWER $\leq$ 0.05. Although we will end up with more false positives, FDR gives us much more power. This makes it particularly appropriate for discovery phase experiments where we may accept FDR levels much higher than 0.05.

6.9   Direct Approach to FDR and q-values (Advanced)

Here we review the results described by John D. Storey in J. R. Statist. Soc. B (2002). One major distinction between Storey's approach and Benjamini and Hochberg's is that we are no longer going to set a α level a priori. Because in many high-throughput experiments we are interested in obtaining some list for validation, we can instead decide beforehand that we will consider all tests with p-values smaller than 0.01. We then want to attach an estimate of an error rate. Using this approach, we are guaranteed to have $R > 0$. Note that in the FDR definition above we assigned $Q = 0$ in the case that $R = V = 0$. We were therefore computing:

$$\text{FDR} = E\left(\frac{V}{R} \mid R > 0\right) \Pr(R > 0)$$

In the approach proposed by Storey, we condition on having a non-empty list, which implies $R > 0$, and we instead compute the *positive FDR*

$$\text{pFDR} = E\left(\frac{V}{R} \mid R > 0\right)$$

A second distinction is that while Benjamini and Hochberg's procedure controls under the worst case scenario, in which all null hypotheses are true ($m = m_0$), Storey proposes that we actually try to estimate m_0 from the data. Because in high-throughput experiments we have so much data, this is certainly possible. The general idea is to pick a relatively high value p-value cut-off, call it λ, and assume that tests obtaining p-values $> \lambda$ are mostly from cases in which the null hypothesis holds. We can then estimate $\pi_0 = m_0/m$ as:

$$\hat{\pi}_0 = \frac{\#\{p_i > \lambda\}}{(1-\lambda)m}$$

There are more sophisticated procedures than this, but they follow the same general idea. Here is an example setting $\lambda = 0.1$. Using the p-values computed above we have:

```
hist(pvals,breaks=seq(0,1,0.05),freq=FALSE)
lambda = 0.1
pi0=sum(pvals> lambda) /((1-lambda)*m)
abline(h= pi0)

print(pi0) ##this is close to the trye pi0=0.9
```

```
## [1] 0.9311111
```

With this estimate in place we can, for example, alter the Benjamini and Hochberg procedures to select the k to be the largest value so that:

$$\hat{\pi}_0 p_{(i)} \leq \frac{i}{m}\alpha$$

However, instead of doing this, we compute a *q-value* for each test. If a feature resulted in a p-value of p, the q-value is the estimated pFDR for a list of all the features with a p-value at least as small as p.

In R, this can be computed with the **qvalue** function in the **qvalue** package:

```
library(qvalue)
res <- qvalue(pvals)
qvals <- res$qvalues
plot(pvals,qvals)
```

we also obtain the estimate of $\hat{\pi}_0$:

```
res$pi0
```

```
## [1] 0.8813727
```

This function uses a more sophisticated approach at estimating π_0 than what is described above.

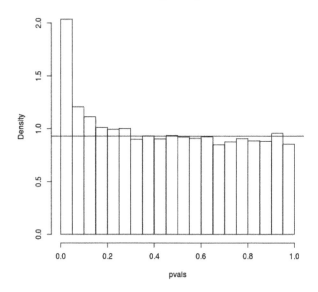

FIGURE 6.9
p-value histogram with pi0 estimate.

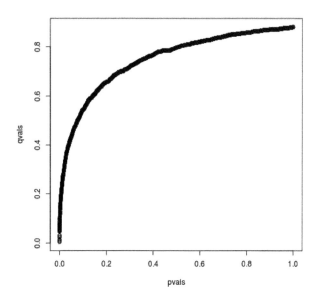

FIGURE 6.10
q-values versus p-values.

Note on estimating π_0 In our experience the estimation of π_0 can be unstable and adds a step of uncertainty to the data analysis pipeline. Although more conservative, the Benjamini-Hochberg procedure is computationally more stable. {pagebreak}

6.10   Exercises

With these exercises we hope to help you further grasp the concept that p-values are random variables and start laying the ground work for the development of procedures that control error rates. The calculations to compute error rates require us to understand the random behavior of p-values.

We are going to ask you to perform some calculations related to introductory probability theory. One particular concept you need to understand is statistical independence. You also will need to know that the probability of two random events that are statistically independent occurring is $P(A \text{ and } B) = P(A)P(B)$. This is a consequence of the more general formula $P(A \text{ and } B) = P(A)P(B|A)$.

1. Assume the null is true and denote the p-value you would get if you ran a test as P. Define the function $f(x) = \Pr(P > x)$. What does $f(x)$ look like?

   (a) A uniform distribution.

   (b) The identity line.

   (c) A constant at 0.05.

   (d) P is not a random value.

2. In the previous exercises, we saw how the probability of incorrectly rejecting the null for at least one of 20 experiments for which the null is true, is well over 5%. Now let's consider a case in which we run thousands of tests, as we would do in a high-throughput experiment.

We previously learned that under the null, the probability of a p-value $<$ p is p. If we run 8,793 independent tests, what it the probability of incorrectly rejecting at least one of the null hypotheses?

3. Suppose we need to run 8,793 statistical tests and we want to make the probability of a mistake very small, say 5%. Use the answer to exercise 2 to determine how small we have to change the cutoff, previously 0.05, to lower our probability of at least one mistake to be 5%.

The following exercises should help you understand the concept of an error controlling procedure. You can think of it as defining a set of instructions, such as "reject all the null hypothesis for which p-values $<$ 0.0001" or "reject the null hypothesis for the 10 features with smallest p-values". Then, knowing the p-values are random variables, we use statistical theory to compute how many mistakes, on average, we will make if we follow this procedure. More precisely, we commonly find bounds on these rates, meaning that we show that they are smaller than some predetermined value.

As described in the text, we can compute different error rates. The FWER tells us the probability of having at least one false positive. The FDR is the expected rate of rejected null hypothesis.

Note 1: the FWER and FDR are not procedures, but error rates. We will review procedures here and use Monte Carlo simulations to estimate their error rates.

Note 2: We sometimes use the colloquial term "pick genes that" meaning "reject the null hypothesis for genes that".

4. We have learned about the family wide error rate FWER. This is the probability of incorrectly rejecting the null at least once. Using the notation introduced in this chapter, this probability is written like this: $\Pr(V > 0)$.

What we want to do in practice is choose a *procedure* that guarantees this probability is smaller than a predetermined value such as 0.05. Here we keep it general and, instead of 0.05, we use α.

We have already learned that the procedure "pick all the genes with p-value <0.05" fails miserably as we have seen that $Pr(V > 0) \approx 1$. So what else can we do?

The Bonferroni procedure assumes we have computed p-values for each test and asks what constant k should we pick so that the procedure "pick all genes with p-value less than k" has $\Pr(V > 0) = 0.05$. Furthermore, we typically want to be conservative rather than lenient, so we accept a procedure that has $\Pr(V > 0) \leq 0.05$.

So the first result we rely on is that this probability is largest when all the null hypotheses are true:

$$\Pr(V > 0) \leq \Pr(V > 0 | \text{all nulls are true})$$

or:

$$\Pr(V > 0) \leq \Pr(V > 0 | m_1 = 0)$$

We showed that if the tests are independent then:

$$\Pr(V > 0 | m_1) = 1 - (1 - k)^m$$

And we pick k so that $1 - (1 - k)^m = \alpha \implies k = 1 - (1 - \alpha)^{1/m}$

Now this requires the tests to be independent. The Bonferroni procedure does not make this assumption and, as we previously saw, sets $k = \alpha/m$ and shows that with this choice of k this procedure results in $Pr(V > 0) \leq \alpha$.

In R define

```
alphas <- seq(0,0.25,0.01)
```

Make a plot of α/m and $1 - (1 - \alpha)^{1/m}$ for various values of m>1.
Which procedure is more conservative Bonferroni's or Sidek's?

(a) They are the same.

(b) Bonferroni's.

(c) Depends on m.

(d) Sidek's.

5. To simulate the p-value results of, say 8,792 t-tests for which the null is true, we don't actually have to generate the original data. We can generate p-values for a uniform distribution like this: `pvals <- runif(8793,0,1)`. Using what we have learned, set the cutoff using the Bonferroni correction and report back the FWER. Set the seed at 1 and run 10,000 simulations.

6. Using the same seed, repeat exercise 5, but for Sidek's cutoff.

7. In the following exercises, we will define error controlling procedures for experimental data. We will make a list of genes based on q-values. We will also assess your understanding of false positives rates and false negative rates by asking you to create a Monte Carlo simulation.

Load the gene expression data:

```
library(GSE5859Subset)
data(GSE5859Subset)
```

We are interested in comparing gene expression between the two groups defined in the `sampleInfo` table.

Compute a p-value for each gene using the function `rowttests` from the genefilter package.

```
library(genefilter)
?rowttests
```

How many genes have p-values smaller than 0.05?

8. Apply the Bonferroni correction to achieve a FWER of 0.05. How many genes are called significant under this procedure?

9. The FDR is a property of a list of features, not each specific feature. The q-value relates FDR to individual features. To define the q-value, we order features we tested by p-value, then compute the FDRs for a list with the most significant, the two most significant, the three most significant, etc. . The FDR of the list with the, say, m most significant tests is defined as the q-value of the m-th most significant feature. In other words, the q-value of a feature is the FDR of the biggest list that includes that gene.

In R, we can compute q-values using the `p.adjust` function with the FDR option. Read the help file for `p.adjust` and compute how many genes achieve a q-value < 0.05 for our gene expression dataset.

10. Now use the `qvalue` function, in the Bioconductor `qvalue` package, to estimate q-values using the procedure described by Storey. How many genes have q-values below 0.05?

11. Read the help file for qvalue and report the estimated proportion of genes for which the null hypothesis is true $\pi_0 = m_0/m$

12. The number of genes passing the q-value <0.05 threshold is larger with the q-value function than the p.adjust difference. Why is this the case? Make a plot of the ratio of these two estimates to help answer the question.

    (a) One of the two procedures is flawed.

    (b) The two functions are estimating different things.

    (c) The qvalue function estimates the proportion of genes for which the null hypothesis is true and provides a less conservative estimate.

    (d) The qvalue function estimates the proportion of genes for which the null hypothesis is true and provides a more conservative estimate.

13. This exercise and the remaining one are more advanced. Create a Monte Carlo simulation in which you simulate measurements from 8,793 genes for 24 samples, 12 cases and 12 controls. For 100 genes, create a difference of 1 between cases and

controls. You can use the code provided below. Run this experiment 1,000 times with a Monte Carlo simulation. For each instance, compute p-values using a t-test and keep track of the number of false positives and false negatives. Compute the false positive rate and false negative rate if we use Bonferroni, q-values from `p.adjust`, and q-values from `qvalue` function. Set the seed to 1 for all three simulations. What is the false positive rate for Bonferroni?

```
n <- 24
m <- 8793
mat <- matrix(rnorm(n*m),m,n)
delta <- 1
positives <- 500
mat[1:positives,1:(n/2)] <- mat[1:positives,1:(n/2)]+delta
```

14. What are the false negative rates for Bonferroni?

15. What are the false positive rates for `p.adjust`?

16. What are the false negative rates for `p.adjust`?

17. What are the false positive rates for `qvalues`?

18. What are the false negative rates for `qvalues`?

6.11 Basic Exploratory Data Analysis

An under-appreciated advantage of working with high-throughput data is that problems with the data are sometimes more easily exposed than with low-throughput data. The fact that we have thousands of measurements permits us to see problems that are not apparent when only a few measurements are available. A powerful way to detect these problems is with exploratory data analysis (EDA). Here we review some of the plots that allow us to detect quality problems.

Volcano plots Here we will use the results obtained from applying t-test to data from a gene expression dataset:

```
library(genefilter)
library(GSE5859Subset)
data(GSE5859Subset)
g <- factor(sampleInfo$group)
results <- rowttests(geneExpression,g)
pvals <- results$p.value
```

And we also generate p-values from a dataset for which we know the null is true:

```
m <- nrow(geneExpression)
n <- ncol(geneExpression)
randomData <- matrix(rnorm(n*m),m,n)
nullpvals <- rowttests(randomData,g)$p.value
```

As we described earlier, reporting only p-values is a mistake when we can also report effect sizes. With high-throughput data, we can visualize the results by making a *volcano plot.* The idea behind a volcano plot is to show these for all features. In the y-axis we plot -log (base 10) p-values and on the x-axis we plot the effect size. By using -log (base 10), the "highly significant" features appear at the top of the plot. Using log also permits us to better distinguish between small and very small p-values, for example 0.01 and 10^6. Here is the volcano plot for our results above:

```
plot(results$dm,-log10(results$p.value),
     xlab="Effect size",ylab="- log (base 10) p-values")
```

Many features with very small p-values, but small effect sizes as we see here, are sometimes indicative of problematic data.

p-value Histograms Another plot we can create to get an overall idea of the results is to make histograms of p-values. When we generate completely null data the histogram follows a uniform distribution. With our original dataset we see a higher frequency of smaller p-values.

```
library(rafalib)
mypar(1,2)
hist(nullpvals,ylim=c(0,1400))
hist(pvals,ylim=c(0,1400))
```

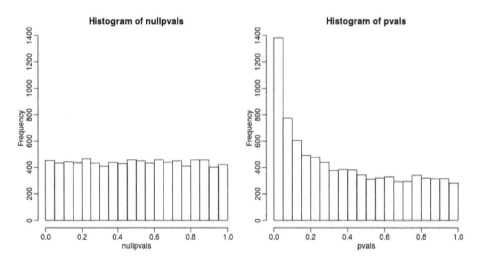

FIGURE 6.11
P-value histogram. We show a simulated case in which all null hypotheses are true (left) and p-values from the gene expression described above.

When we expect most hypotheses to be null and don't see a uniform p-value distribution, it might be indicative of unexpected properties, such as correlated samples.

If we permute the outcomes and calculate p-values then, if the samples are independent, we should see a uniform distribution. With these data we do not:

```
permg <- sample(g)
permresults <- rowttests(geneExpression,permg)
hist(permresults$p.value)
```

In a later chapter we will see that the columns in this dataset are not independent and thus the assumptions used to compute the p-values here are incorrect.

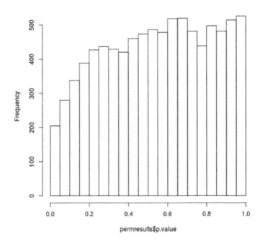

FIGURE 6.12
Histogram obtained after permuting labels.

Data boxplots and histograms With high-throughput data, we have thousands of measurements for each experimental unit. As mentioned earlier, this can help us detect quality issues. For example, if one sample has a completely different distribution than the rest, we might suspect there are problems. Although a complete change in distribution could be due to real biological differences, more often than not it is due to a technical problem. Here we load a large gene expression experiment available from Bioconductor. We "accidentally" use log instead of log2 on one of the samples.

```
library(Biobase)
library(GSE5859)
data(GSE5859)
ge <- exprs(e) ##ge for gene expression
ge[,49] <- ge[,49]/log2(exp(1)) ##imitate error
```

A quick look at a summary of the distribution using boxplots immediately highlights the mistake:

```
library(rafalib)
mypar(1,1)
boxplot(ge,range=0,names=1:ncol(e),col=ifelse(1:ncol(ge)==49,1,2))
```

Note that the number of samples is a bit too large here, making it hard to see the boxes. One can instead simply show the boxplot summaries without the boxes:

```
qs <- t(apply(ge,2,quantile,prob=c(0.05,0.25,0.5,0.75,0.95)))
matplot(qs,type="l",lty=1)
```

We refer to this figure as a *kaboxplot* because Karl Broman was the first we saw use it as an alternative to boxplots.

We can also plot all the histograms. Because we have so much data, we create histograms using small bins, then smooth the heights of the bars and then plot *smooth histograms*. We

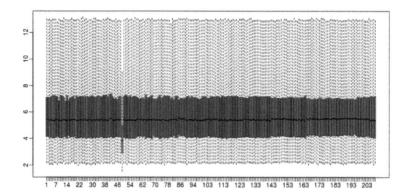

FIGURE 6.13
Boxplot for log-scale expression for all samples.

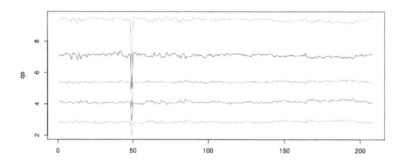

FIGURE 6.14
The 0.05, 0.25, 0.5, 0.75, and 0.95 quantiles are plotted for each sample.

re-calibrate the height of these smooth curves so that if a bar is made with base of size "unit" and height given by the curve at x_0, the area approximates the number of points in region of size "unit" centered at x_0:

```
mypar(1,1)
shist(ge,unit=0.5)
```

MA plot Scatterplots and correlation are not the best tools to detect replication problems. A better measure of replication can be obtained from examining the differences between the values that should be the same. Therefore, a better plot is a rotation of the scatterplot containing the differences on the y-axis and the averages on the x-axis. This plot was originally named a Bland-Altman plot, but in genomics it is commonly referred to as an MA-plot. The name MA comes from plots of red log intensity minus (M) green intensities versus average (A) log intensities used with microarrays (MA) data.

```
x <- ge[,1]
y <- ge[,2]
```

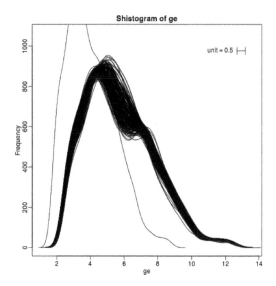

FIGURE 6.15
Smooth histograms for each sample.

```
mypar(1,2)
plot(x,y)
plot((x+y)/2,x-y)
```

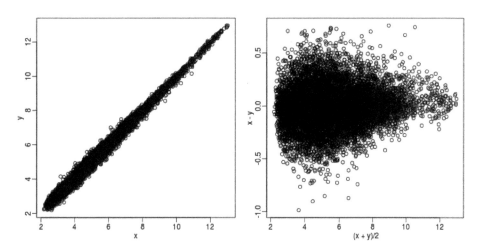

FIGURE 6.16
Scatter plot (left) and M versus A plot (right) for the same data.

Note that once we rotate the plot, the fact that these data have differences of about:

```
sd(y-x)
```

```
## [1] 0.2025465
```

becomes immediate. The scatterplot shows very strong correlation, which is not necessarily informative here.

We will later introduce dendograms, heatmaps, and multi-dimensional scaling plots. {pagebreak}

6.12   Exercises

We will be using a handful of Bioconductor packages. These are installed using the function `biocLite` which you can source from the web:

```
source("http://www.bioconductor.org/biocLite.R")
```

or you can run the `bioc_install` in the `rafalib` package.

```
library(rafalib)
bioc_install()
```

Download and install the Bioconductor package `SpikeInSubset` and then load the library and the `mas133` data:

```
library(rafalib)
install_bioc("SpikeInSubset")
library(SpikeInSubset)
data(mas133)
```

Now make the following plot of the first two samples and compute the correlation:

```
e <- exprs(mas133)
plot(e[,1],e[,2],main=paste0("corr=",signif(cor(e[,1],e[,2]),3)),cex=0.5)
k <- 3000
b <- 1000 #a buffer
polygon(c(-b,k,k,-b),c(-b,-b,k,k),col="red",density=0,border="red")
```

1. What proportion of the points are inside the box?

2. Now make the sample plot with log:

```
plot(log2(e[,1]),log2(e[,2]))
k <- log2(3000)
b <- log2(0.5)
polygon(c(b,k,k,b),c(b,b,k,k),col="red",density=0,border="red")
```

What is an advantage of taking the log?

- A) The tails do not dominate the plot: 95% of data is not in a tiny section of plot.

- B) There are fewer points.

- C) There is exponential growth.

- D) We always take logs.

3. Make an MA-plot:

```
e <- log2(exprs(mas133))
plot((e[,1]+e[,2])/2,e[,2]-e[,1],cex=0.5)
```

The two samples we are plotting are replicates (they are random samples from the same batch of RNA). The correlation of the data was 0.997 in the original scale and 0.96 in the log-scale. High correlations are sometimes confused with evidence of replication. However, replication implies we get very small differences between the observations, which is better measured with distance or differences.

What is the standard deviation of the log ratios for this comparison?

4. How many fold changes above 2 do we see?

7

Statistical Models

All models are wrong, but some are useful. -George E. P. Box

When we see a p-value in the literature, it means a probability distribution of some sort was used to quantify the null hypothesis. Many times deciding which probability distribution to use is relatively straightforward. For example, in the tea tasting challenge we can use simple probability calculations to determine the null distribution. Most p-values in the scientific literature are based on sample averages or least squares estimates from a linear model and make use of the CLT to approximate the null distribution of their statistic as normal.

The CLT is backed by theoretical results that guarantee that the approximation is accurate. However, we cannot always use this approximation, such as when our sample size is too small. Previously, we described how the sample average can be approximated as t-distributed when the population data is approximately normal. However, there is no theoretical backing for this assumption. We are now *modeling*. In the case of height, we know from experience that this turns out to be a very good model.

But this does not imply that every dataset we collect will follow a normal distribution. Some examples are: coin tosses, the number of people who win the lottery, and US incomes. The normal distribution is not the only parametric distribution that is available for modeling. Here we provide a very brief introduction to some of the most widely used parametric distributions and some of their uses in the life sciences. We focus on the models and concepts needed to understand the techniques currently used to perform statistical inference on high-throughput data. To do this we also need to introduce the basics of Bayesian statistics. For more in-depth description of probability models and parametric distributions please consult a Statistics textbook such as this one[1].

7.1 The Binomial Distribution

The first distribution we will describe is the binomial distribution. It reports the probability of observing $S = k$ successes in N trials as

$$\Pr(S = k) = \binom{N}{k} p^k (1 - p)^{N-k}$$

with p the probability of success. The best known example is coin tosses with S the number of heads when tossing N coins. In this example $p = 0.5$.

Note that S/N is the average of independent random variables and thus the CLT tells us that S is approximately normal when N is large. This distribution has many applications in the life sciences. Recently, it has been used by the variant callers and genotypers applied

---

[1]https://www.stat.berkeley.edu/~rice/Book3ed/index.html

to next generation sequencing. A special case of this distribution is approximated by the Poisson distribution which we describe next.

7.2   The Poisson Distribution

Since it is the sum of binary outcomes, the number of people that win the lottery follows a binomial distribution (we assume each person buys one ticket). The number of trials N is the number of people that buy tickets and is usually very large. However, the number of people that win the lottery oscillates between 0 and 3, which implies the normal approximation does not hold. So why does CLT not hold? One can explain this mathematically, but the intuition is that with the sum of successes so close to and also constrained to be larger than 0, it is impossible for the distribution to be normal. Here is a quick simulation:

```
p=10^-7 ##1 in 10,000,0000 chances of winning
N=5*10^6 ##5,000,000 tickets bought
winners=rbinom(1000,N,p) ##1000 is the number of different lotto draws
tab=table(winners)
plot(tab)
```

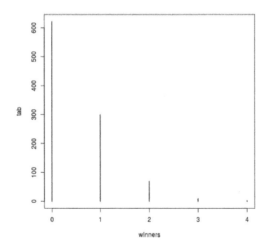

FIGURE 7.1
Number of people that win the lottery obtained from Monte Carlo simulation.

```
prop.table(tab)
```

```
## winners
##     0     1     2     3     4
## 0.621 0.299 0.069 0.009 0.002
```

For cases like this, where N is very large, but p is small enough to make $N \times p$ (call it λ) a number between 0 and, for example, 10, then S can be shown to follow a Poisson distribution, which has a simple parametric form:

$$\Pr(S = k) = \frac{\lambda^k \exp{-\lambda}}{k!}$$

The Poisson distribution is commonly used in RNA-seq analyses. Because we are sampling thousands of molecules and most genes represent a very small proportion of the totality of molecules, the Poisson distribution seems appropriate.

So how does this help us? One way is that it provides insight about the statistical properties of summaries that are widely used in practice. For example, let's say we only have one sample from each of a case and control RNA-seq experiment and we want to report the genes with the largest fold-changes. One insight that the Poisson model provides is that under the null hypothesis of no true differences, the statistical variability of this quantity depends on the total abundance of the gene. We can show this mathematically, but here is a quick simulation to demonstrate the point:

```
N=10000##number of genes
lambdas=2^seq(1,16,len=N) ##these are the true abundances of genes
y=rpois(N,lambdas)##note that the null hypothesis is true for all genes
x=rpois(N,lambdas)
ind=which(y>0 & x>0)##make sure no 0s due to ratio and log

library(rafalib)
splot(log2(lambdas),log2(y/x),subset=ind)
```

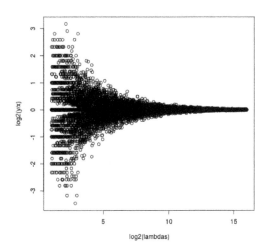

FIGURE 7.2
MA plot of simulated RNA-seq data. Replicated measurements follow a Poisson distribution.

For lower values of `lambda` there is much more variability and, if we were to report anything with a fold change of 2 or more, the number of false positives would be quite high for low abundance genes.

NGS experiments and the Poisson distribution In this section we will use the data stored in this dataset:

```
library(parathyroidSE) ##available from Bioconductor
data(parathyroidGenesSE)
se <- parathyroidGenesSE
```

The data is contained in a `SummarizedExperiment` object, which we do not describe here. The important thing to know is that it includes a matrix of data, where each row is a genomic feature and each column is a sample. We can extract this data using the `assay` function. For this dataset, the value of a single cell in the data matrix is the count of reads which align to a given gene for a given sample. Thus, a similar plot to the one we simulated above with technical replicates reveals that the behavior predicted by the model is present in experimental data:

```
x <- assay(se)[,23]
y <- assay(se)[,24]
ind=which(y>0 & x>0)##make sure no 0s due to ratio and log
splot((log2(x)+log2(y))/2,log(x/y),subset=ind)
```

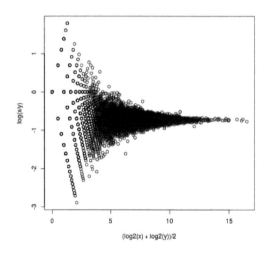

FIGURE 7.3
MA plot of replicated RNA-seq data.

If we compute the standard deviations across four individuals, it is quite a bit higher than what is predicted by a Poisson model. Assuming most genes are differentially expressed across individuals, then, if the Poisson model is appropriate, there should be a linear relationship in this plot:

```
library(rafalib)
library(matrixStats)

vars=rowVars(assay(se)[,c(2,8,16,21)]) ##we know these are four
means=rowMeans(assay(se)[,c(2,8,16,21)]) ##different individuals

splot(means,vars,log="xy",subset=which(means>0&vars>0)) ##plot a subset of data
abline(0,1,col=2,lwd=2)
```

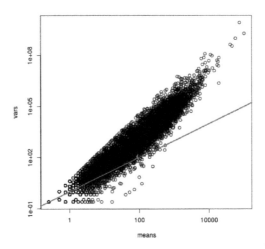

FIGURE 7.4
Variance versus mean plot. Summaries were obtained from the RNA-seq data.

The reason for this is that the variability plotted here includes biological variability, which the motivation for the Poisson does not include. The negative binomial distribution, which combines the sampling variability of a Poisson and biological variability, is a more appropriate distribution to model this type of experiment. The negative binomial has two parameters and permits more flexibility for count data. For more on the use of the negative binomial to model RNA-seq data you can read this paper[2]. The Poisson is a special case of the negative binomial distribution.

7.3   Maximum Likelihood Estimation

To illustrate the concept of maximum likelihood estimates (MLE), we use a relatively simple dataset containing palindrome locations in the HMCV genome. We read in the locations of the palindrome and then count the number of palindromes in each 4,000 basepair segments.

```
datadir="http://www.biostat.jhsph.edu/bstcourse/bio751/data"
x=read.csv(file.path(datadir,"hcmv.csv"))[,2]

breaks=seq(0,4000*round(max(x)/4000),4000)
tmp=cut(x,breaks)
counts=table(tmp)

library(rafalib)
mypar(1,1)
hist(counts)
```

The counts do appear to follow a Poisson distribution. But what is the rate λ ? The

[2]http://www.ncbi.nlm.nih.gov/pubmed/20979621

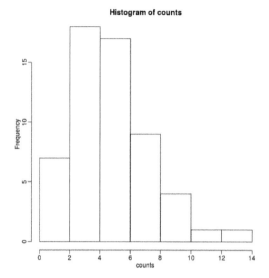

FIGURE 7.5
Palindrome count histogram.

most common approach to estimating this rate is *maximum likelihood estimation*. To find the maximum likelihood estimate (MLE), we note that these data are independent and the probability of observing the values we observed is:

$$\Pr(X_1 = k_1, \ldots, X_n = k_n; \lambda) = \prod_{i=1}^{n} \lambda^{k_i} / k_i! \exp(-\lambda)$$

The MLE is the value of λ that maximizes the likelihood:.

$$L(\lambda; X_1 = k_1, \ldots, X_n = k_1) = \exp \left\{ \sum_{i=1}^{n} \log \Pr(X_i = k_i; \lambda) \right\}$$

In practice, it is more convenient to maximize the log-likelihood which is the summation that is exponentiated in the expression above. Below we write code that computes the log-likelihood for any λ and use the function `optimize` to find the value that maximizes this function (the MLE). We show a plot of the log-likelihood along with vertical line showing the MLE.

```
l<-function(lambda) sum(dpois(counts,lambda,log=TRUE))

lambdas<-seq(3,7,len=100)
ls <- exp(sapply(lambdas,l))

plot(lambdas,ls,type="l")

mle=optimize(l,c(0,10),maximum=TRUE)
abline(v=mle$maximum)
```

If you work out the math and do a bit of calculus, you realize that this is a particularly simple example for which the MLE is the average.

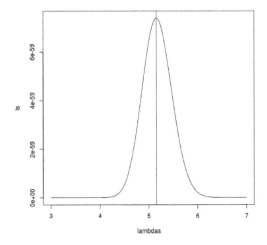

FIGURE 7.6
Likelihood versus lambda.

```
print( c(mle$maximum, mean(counts) ) )
```

```
## [1] 5.157894 5.157895
```

Note that a plot of observed counts versus counts predicted by the Poisson shows that the fit is quite good in this case:

```
theoretical<-qpois((seq(0,99)+0.5)/100,mean(counts))
```

```
qqplot(theoretical,counts)
abline(0,1)
```

We therefore can model the palindrome count data with a Poisson with $\lambda = 5.16$.

7.4 Distributions for Positive Continuous Values

Different genes vary differently across biological replicates. Later, in the hierarchical models chapter, we will describe one of the most influential statistical methods[3] in the analysis of genomics data. This method provides great improvements over naive approaches to detecting differentially expressed genes. This is achieved by modeling the distribution of the gene variances. Here we describe the parametric model used in this method.

We want to model the distribution of the gene-specific standard errors. Are they normal? Keep in mind that we are modeling the population standard errors so CLT does not apply, even though we have thousands of genes.

As an example, we use an experimental data that included both technical and biological replicates for gene expression measurements on mice. We can load the data and compute

---

[3]http://www.ncbi.nlm.nih.gov/pubmed/16646809

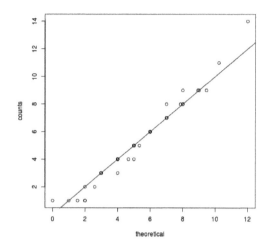

FIGURE 7.7
Observed counts versus theoretical Poisson counts.

the gene specific sample standard error for both the technical replicates and the biological replicates

```
library(Biobase) ##available from Bioconductor
library(maPooling) ##available from course github repo

data(maPooling)
pd=pData(maPooling)

##determine which samples are bio reps and which are tech reps
strain=factor(as.numeric(grepl("b",rownames(pd))))
pooled=which(rowSums(pd)==12 & strain==1)
techreps=exprs(maPooling[,pooled])
individuals=which(rowSums(pd)==1 & strain==1)

##remove replicates
individuals=individuals[-grep("tr",names(individuals))]
bioreps=exprs(maPooling)[,individuals]

###now compute the gene specific standard deviations
library(matrixStats)
techsds=rowSds(techreps)
biosds=rowSds(bioreps)
```

We can now explore the sample standard deviation:

```
###now plot
library(rafalib)
mypar()
shist(biosds,unit=0.1,col=1,xlim=c(0,1.5))
```

```
shist(techsds,unit=0.1,col=2,add=TRUE)
legend("topright",c("Biological","Technical"), col=c(1,2),lty=c(1,1))
```

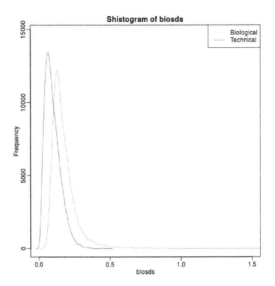

FIGURE 7.8
Histograms of biological variance and technical variance.

An important observation here is that the biological variability is substantially higher than the technical variability. This provides strong evidence that genes do in fact have gene-specific biological variability.

If we want to model this variability, we first notice that the normal distribution is not appropriate here since the right tail is rather large. Also, because SDs are strictly positive, there is a limitation to how symmetric this distribution can be. A qqplot shows this very clearly:

```
qqnorm(biosds)
qqline(biosds)
```

There are parametric distributions that possess these properties (strictly positive and *heavy* right tails). Two examples are the *gamma* and *F* distributions. The density of the gamma distribution is defined by:

$$f(x; \alpha, \beta) = \frac{\beta^{\alpha} x^{\alpha-1} \exp{-\beta x}}{\Gamma(\alpha)}$$

It is defined by two parameters α and β that can, indirectly, control location and scale. They also control the shape of the distribution. For more on this distribution please refer to this book[4].

Two special cases of the gamma distribution are the chi-squared and exponential distribution. We used the chi-squared earlier to analyze a 2x2 table data. For chi-square, we have $\alpha = \nu/2$ and $\beta = 2$ with ν the degrees of freedom. For exponential, we have $\alpha = 1$ and $\beta = \lambda$ the rate.

The F-distribution comes up in analysis of variance (ANOVA). It is also always positive and has large right tails. Two parameters control its shape:

---

[4]https://www.stat.berkeley.edu/~rice/Book3ed/index.html

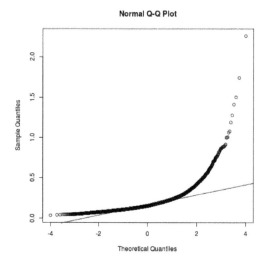

FIGURE 7.9
Normal qq-plot for sample standard deviations.

$$f(x, d_1, d_2) = \frac{1}{B\left(\frac{d_1}{2}, \frac{d_2}{2}\right)} \left(\frac{d_1}{d_2}\right)^{\frac{d_1}{2}} x^{\frac{d_1}{2}-1} \left(1 + \frac{d1}{d2}x\right)^{-\frac{d_1+d_2}{2}}$$

with B the *beta function* and d_1 and d_2 are called the degrees of freedom for reasons having to do with how it arises in ANOVA. A third parameter is sometimes used with the F-distribution, which is a scale parameter.

Modeling the variance In a later section we will learn about a hierarchical model approach to improve estimates of variance. In these cases it is mathematically convenient to model the distribution of the variance σ^2. The hierarchical model used here[5] implies that the sample standard deviation of genes follows scaled F-statistics:

$$s^2 \sim s_0^2 F_{d,d_0}$$

with d the degrees of freedom involved in computing s^2. For example, in a case comparing 3 versus 3, the degrees of freedom would be 4. This leaves two free parameters to adjust to the data. Here d will control the location and s_0 will control the scale. Below are some examples of F distribution plotted on top of the histogram from the sample variances:

```
library(rafalib)
mypar(3,3)
sds=seq(0,2,len=100)
for(d in c(1,5,10)){
  for(s0 in c(0.1, 0.2, 0.3)){
    tmp=hist(biosds,main=paste("s_0 =",s0,"d =",d),
      xlab="sd",ylab="density",freq=FALSE,nc=100,xlim=c(0,1))
    dd=df(sds^2/s0^2,11,d)
    ##multiply by normalizing constant to assure same range on plot
```

---

[5]http://www.ncbi.nlm.nih.gov/pubmed/16646809

```
    k=sum(tmp$density)/sum(dd)
    lines(sds,dd*k,type="l",col=2,lwd=2)
    }
}
```

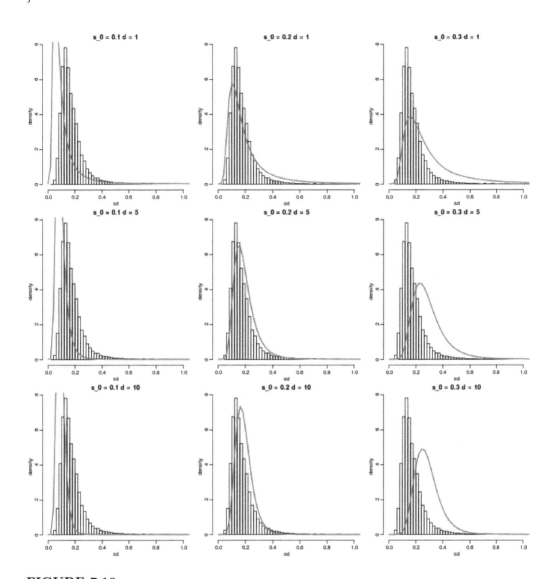

FIGURE 7.10
Histograms of sample standard deviations and densities of estimated distributions.

Now which s_0 and d fit our data best? This is a rather advanced topic as the MLE does not perform well for this particular distribution (we refer to Smyth (2004)). The Bioconductor `limma` package provides a function to estimate these parameters:

```
library(limma)
estimates=fitFDist(biosds^2,11)

theoretical<- sqrt(qf((seq(0,999)+0.5)/1000, 11, estimates$df2)*estimates$scale)
observed <- biosds
```

The fitted models do appear to provide a reasonable approximation, as demonstrated by the qq-plot and histogram:

```
mypar(1,2)
qqplot(theoretical,observed)
abline(0,1)
tmp=hist(biosds,main=paste("s_0 =", signif(estimates[[1]],2),
                "d =", signif(estimates[[2]],2)),
   xlab="sd", ylab="density", freq=FALSE, nc=100, xlim=c(0,1), ylim=c(0,9))
dd=df(sds^2/estimates$scale,11,estimates$df2)
k=sum(tmp$density)/sum(dd) ##a normalizing constant to assure same area in plot
lines(sds, dd*k, type="l", col=2, lwd=2)
```

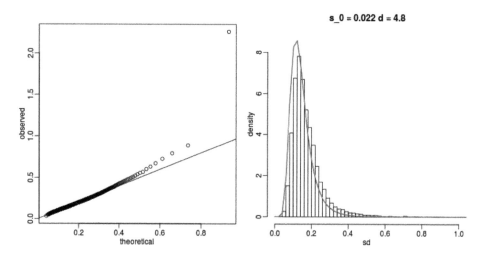

FIGURE 7.11
qq-plot (left) and density (right) demonstrate that model fits data well.

7.5   Exercises

1. Suppose you have an urn with blue and red balls. If N balls are selected at random with replacement (you put the ball back after you pick it), we can denote the outcomes as random variables $X_1, \ldots, X_N$ that are 1 or 0. If the proportion of red balls is p, then the distribution of each of these is $\Pr(X_i = 1) = p$.

These are also called Bernoulli trials. These random variables are independent because we replace the balls. Flipping a coin is an example of this with $p = 0.5$.

You can show that the mean and variance are p and $p(1 - p)$ respectively. The binomial distribution gives us the distribution of the sum S_N of these random variables. The probability that we see k red balls is given by:

$$\Pr(S_N = k) = \binom{N}{k} p^k (1 - p)^{N-k}$$

In R, the function `dbimom` gives you this result. The function `pbinom` gives us $\Pr(S_N \leq k)$.

This equation has many uses in the life sciences. We give some examples below.

The probability of conceiving a girl is 0.49. What is the probability that a family with 4 children has 2 girls and 2 boys (you can assume that the outcomes are independent)?

2. Use what you learned in Question 1 to answer these questions:

What is the probability that a family with 10 children has 6 girls and 4 boys (assume no twins)?

3. The genome has 3 billion bases. About 20% are C, 20% are G, 30% are T, and 30% are A. Suppose you take a random interval of 20 bases, what is the probability that the GC-content (proportion of Gs of Cs) is strictly above 0.5 in this interval?

4. The probability of winning the lottery is 1 in 175,223,510. If 20,000,000 people buy a ticket, what is the probability that more than one person wins?

5. We can show that the binomial approximation is approximately normal when N is large and p is not too close to 0 or 1. This means that:

$$\frac{S_N - \mathrm{E}(S_N)}{\sqrt{\mathrm{Var}(S_N)}}$$

is approximately normal with mean 0 and SD 1. Using the results for sums of independent random variables, we can show that $\mathrm{E}(S_N) = Np$ and $\mathrm{Var}(S_n) = Np(1-p)$.

The genome has 3 billion bases. About 20% are C, 20% are G, 30% are T, and 30% are A. Suppose you take a random interval of 20 bases, what is the exact probability that the GC-content (proportion of Gs of Cs) is greater than 0.35 and smaller or equal to 0.45 in this interval?

6. For the question above, what is the normal approximation to the probability?

7. Repeat exercise 4, but using an interval of 1000 bases. What is the difference (in absolute value) between the normal approximation and the exact distribution of the GC-content being greater than 0.35 and lesser or equal to 0.45?

8. The Cs in our genomes can be *methylated*[6] or *unmethylated*. Suppose we have a large (millions) group of cells in which a proportion p of the Cs of interest are methylated. We break up the DNA of these cells and randomly select pieces and end up with N pieces that contain the C we care about. This means that the probability of seeing k methylated Cs is binomial:

```
exact = dbinom(k,N,p)
```

We can approximate this with the normal distribution:

```
a <- (k+0.5 - N*p)/sqrt(N*p*(1-p))
b <- (k-0.5 - N*p)/sqrt(N*p*(1-p))
approx = pnorm(a) - pnorm(b)
```

Compute the difference `approx - exact` for:

```
N <- c(5,10,50,100,500)
p <- seq(0,1,0.25)
```

---

[6]http://en.wikipedia.org/wiki/DNA_methylation

Compare the approximation and exact probability of the proportion of Cs being p, $k = 1, \ldots, N-1$ plotting the exact versus the approximation for each p and N combination.

- A) The normal approximation works well when p is close to 0.5 even for small $N = 10$

- B) The normal approximation breaks down when p is close to 0 or 1 even for large N

- C) When N is 100 all approximations are spot on.

- D) When $p = 0.01$ the approximations are terrible for $N = 5, 10, 30$ and only OK for $N = 100$

9. We saw in the previous question that when p is very small, the normal approximation breaks down. If N is very large, then we can use the Poisson approximation.

Earlier we computed 1 or more people winning the lottery when the probability of winning was 1 in 175,223,510 and 20,000,000 people bought a ticket. Using the binomial, we can compute the probability of exactly two people winning to be:

```
N <- 20000000
p <- 1/175223510
dbinom(2,N,p)
```

If we were to use the normal approximation, we would greatly underestimate this:

```
a <- (2+0.5 - N*p)/sqrt(N*p*(1-p))
b <- (2-0.5 - N*p)/sqrt(N*p*(1-p))
pnorm(a) - pnorm(b)
```

To use the Poisson approximation here, use the rate $\lambda = Np$ representing the number of people per 20,000,000 that win the lottery. Note how much better the approximation is:

```
dpois(2,N*p)
```

In this case, it is practically the same because N is very large and Np is not 0. These are the assumptions needed for the Poisson to work. What is the Poisson approximation for more than one person winning?

10. Now we are going to explore if palindromes are over-represented in some part of the HCMV genome. Make sure you have the latest version of the **dagdata**, load the palindrome data from the human cytomegalovirus genome, and plot locations of palindromes on the genome for this virus:

```
library(dagdata)
data(hcmv)
plot(locations,rep(1,length(locations)),ylab="",yaxt="n")
```

These palindromes are quite rare, and therefore p is very small. If we break the genome into bins of 4000 basepairs, then we have Np not so small and we might be able to use Poisson to model the number of palindromes in each bin:

```
breaks=seq(0,4000*round(max(locations)/4000),4000)
tmp=cut(locations,breaks)
counts=as.numeric(table(tmp))
```

So if our model is correct, `counts` should follow a Poisson distribution. The distribution seems about right:

```
hist(counts)
```

So let $X_1, \ldots, X_n$ be the random variables representing counts then $\Pr(X_i = k) = \lambda^k/k! \exp(-\lambda)$ and to fully describe this distribution, we need to know λ. For this we will use MLE. We can write the likelihood as a product of probabilities. For example, for $\lambda = 4$ we have:

```
probs <- dpois(counts,4)
likelihood <- prod(probs)
likelihood
```

Notice that it's a tiny number. It is usually more convenient to compute log-likelihoods:

```
logprobs <- dpois(counts,4,log=TRUE)
loglikelihood <- sum(logprobs)
loglikelihood
```

Now write a function that takes λ and the vector of counts as input and returns the log-likelihood. Compute this log-likelihood for `lambdas = seq(0,15,len=300)` and make a plot. What value of `lambdas` maximizes the log-likelihood?

11. The point of collecting this dataset was to try to determine if there is a region of the genome that has a higher palindrome rate than expected. We can create a plot and see the counts per location:

```
library(dagdata)
data(hcmv)
breaks=seq(0,4000*round(max(locations)/4000),4000)
tmp=cut(locations,breaks)
counts=as.numeric(table(tmp))
binLocation=(breaks[-1]+breaks[-length(breaks)])/2
plot(binLocation,counts,type="l",xlab=)
```

What is the center of the bin with the highest count?

12. What is the maximum count?

13. Once we have identified the location with the largest palindrome count, we want to know if we could see a value this big by chance. If X is a Poisson random variable with rate:

```
lambda = mean(counts[ - which.max(counts) ])
```

What is the probability of seeing a count of 14 or more?

14. So we obtain a p-value smaller than 0.001 for a count of 14. Why is it problematic to report this p-value as strong evidence of a location that is different?
    (a) Poisson in only an approximation.
    (b) We selected the highest region out of 57 and need to adjust for multiple testing.
    (c) λ is an estimate, a random variable, and we didn't take into account its variability.

    (d) We don't know the effect size.

15. Use the Bonferroni correction to determine the p-value cut-off that guarantees a FWER of 0.05. What is this p-value cutoff?

16. Create a qq-plot to see if our Poisson model is a good fit:

```
ps <- (seq(along=counts) - 0.5)/length(counts)
lambda <- mean( counts[ -which.max(counts)])
poisq <- qpois(ps,lambda)
plot(poisq,sort(counts))
abline(0,1)
```

How would you characterize this qq-plot - A) Poisson is a terrible approximation. - B) Poisson is a very good approximation except for one point that we actually think is a region of interest. - C) There are too many 1s in the data. - D) A normal distribution provides a better approximation.

17. Load the `tissuesGeneExpression` data library

```
library(tissuesGeneExpression)
```

Now load this data and select the columns related to endometrium:

```
library(genefilter)
y = e[,which(tissue=="endometrium")]
```

This will give you a matrix y with 15 samples. Compute the across sample variance for the first three samples. Then make a qq-plot to see if the data follow a normal distribution. Which of the following is true? - A) With the exception of a handful of outliers, the data follow a normal distribution. - B) The variance does not follow a normal distribution, but taking the square root fixes this. - C) The normal distribution is not usable here: the left tail is over estimated and the right tail is underestimated. - D) The normal distribution fits the data almost perfectly.

18. Now fit an F-distribution with 14 degrees of freedom using the `fitFDist` function in the `limma` package.

19. Now create a qq-plot of the observed sample variances versus the F-distribution quantiles. Which of the following best describes the qq-plot?

    (a) The fitted F-distribution provides a perfect fit.

    (b) If we exclude the lowest 0.1% of the data, the F-distribution provides a good fit.

    (c) The normal distribution provided a better fit.

    (d) If we exclude the highest 0.1% of the data, the F-distribution provides a good fit.

7.6 Bayesian Statistics

One distinguishing characteristic of high-throughput data is that although we want to report on specific features, we observe many related outcomes. For example, we measure the expression of thousands of genes, or the height of thousands of peaks representing protein binding, or the methylation levels across several CpGs. However, most of the statistical inference approaches we have shown here treat each feature independently and pretty much ignores data from other features. We will learn how using statistical models provides power by modeling features jointly. The most successful of these approaches are what we refer to as hierarchical models, which we explain below in the context of Bayesian statistics.

Bayes theorem We start by reviewing Bayes theorem. We do this using a hypothetical cystic fibrosis test as an example. Suppose a test for cystic fibrosis has an accuracy of 99%. We will use the following notation:

$$\text{Prob}(+ \mid D = 1) = 0.99, \text{Prob}(- \mid D = 0) = 0.99$$

with $+$ meaning a positive test and D representing if you actually have the disease (1) or not (0).

Suppose we select a random person and they test positive, what is the probability that they have the disease? We write this as $\text{Prob}(D = 1 \mid +)$? The cystic fibrosis rate is 1 in 3,900 which implies that $\text{Prob}(D = 1) = 0.00025$. To answer this question we will use Bayes Theorem, which in general tells us that:

$$\text{Pr}(A \mid B) = \frac{\text{Pr}(B \mid A)\text{Pr}(A)}{\text{Pr}(B)}$$

This equation applied to our problem becomes:

$$\text{Prob}(D = 1 \mid +) = \frac{P(+ \mid D = 1) \cdot P(D = 1)}{\text{Prob}(+)}$$

$$= \frac{\text{Prob}(+ \mid D = 1) \cdot P(D = 1)}{\text{Prob}(+ \mid D = 1) \cdot P(D = 1) + \text{Prob}(+ \mid D = 0)\text{Prob}(D = 0)}$$

Plugging in the numbers we get:

$$\frac{0.99 \cdot 0.00025}{0.99 \cdot 0.00025 + 0.01 \cdot (.99975)} = 0.02$$

This says that despite the test having 0.99 accuracy, the probability of having the disease given a positive test is only 0.02. This may appear counterintuitive to some. The reason this is the case is because we have to factor in the very rare probability that a person, chosen at random, has the disease. To illustrate this we run a Monte Carlo simulation.

Simulation The following simulation is meant to help you visualize Bayes Theorem. We start by randomly selecting 1500 people from a population in which the disease in question has a 5% prevalence.

```
set.seed(3)
prev <- 1/20
##Later, we are arranging 1000 people in 80 rows and 20 columns
M <- 50 ; N <- 30
##do they have the disease?
d<-rbinom(N*M,1,p=prev)
```

Now each person gets the test which is correct 90% of the time.

```
accuracy <- 0.9
test <- rep(NA,N*M)
##do controls test positive?
test[d==1]  <- rbinom(sum(d==1), 1, p=accuracy)
##do cases test positive?
test[d==0]  <- rbinom(sum(d==0), 1, p=1-accuracy)
```

Because there are so many more controls than cases, even with a low false positive rate, we get more controls than cases in the group that tested positive (code not shown):

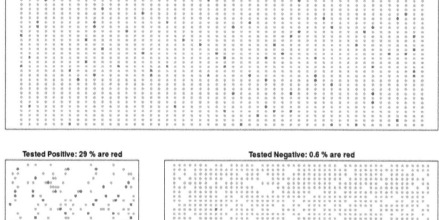

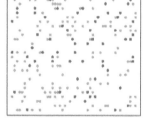

FIGURE 7.12
Simulation demonstrating Bayes theorem. Top plot shows every individual with red denoting cases. Each one takes a test and with 90% gives correct answer. Those called positive (either correctly or incorrectly) are put in the bottom left pane. Those called negative in the bottom right.

The proportions of red in the top plot shows $\Pr(D = 1)$. The bottom left shows $\Pr(D = 1 \mid +)$ and the bottom right shows $\Pr(D = 0 \mid +)$.

Bayes in practice Jos Iglesias is a professional baseball player. In April 2013, when he was starting his career, he was performing rather well:

Month	At Bats	H	AVG
April	20	9	.450

The batting average (`AVG`) statistic is one way of measuring success. Roughly speaking, it tells us the success rate when batting. An `AVG` of .450 means Jos has been successful 45% of the times he has batted (`At Bats`) which is rather high as we will see. Note, for example, that no one has finished a season with an `AVG` of .400 since Ted Williams did it in 1941! To illustrate the way hierarchical models are powerful, we will try to predict Jos's batting average at the end of the season, after he has gone to bat over 500 times.

With the techniques we have learned up to now, referred to as *frequentist techniques*, the best we can do is provide a confidence interval. We can think of outcomes from hitting as a binomial with a success rate of p. So if the success rate is indeed .450, the standard error of just 20 at bats is:

$$\sqrt{\frac{.450(1 - .450)}{20}} = .111$$

This means that our confidence interval is .450-.222 to .450+.222 or .228 to .672.

This prediction has two problems. First, it is very large so not very useful. Also, it is centered at .450 which implies that our best guess is that this new player will break Ted Williams' record. If you follow baseball, this last statement will seem wrong and this is because you are implicitly using a hierarchical model that factors in information from years of following baseball. Here we show how we can quantify this intuition.

First, let's explore the distribution of batting averages for all players with more than 500 at bats during the previous three seasons:

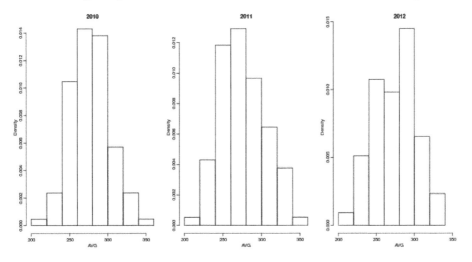

FIGURE 7.13
Batting average histograms for 2010, 2011, and 2012.

We note that the average player had an `AVG` of .275 and the standard deviation of the population of players was 0.027. So we can see already that .450 would be quite an anomaly since it is over six SDs away from the mean. So is Jos lucky or the best batter seen in the last 50 years? Perhaps it's a combination of both. But how lucky and how good is he? If we become convinced that he is lucky, we should trade him to a team that trusts the .450 observation and is maybe overestimating his potential.

The hierarchical model The hierarchical model provides a mathematical description of how we came to see the observation of .450. First, we pick a player at random with an intrinsic ability summarized by, for example, θ, then we see 20 random outcomes with success probability θ.

$\theta \sim N(\mu, \tau^2)$ describes randomness in picking a player

$Y \mid \theta \sim N(\theta, \sigma^2)$ describes randomness in the performance of this particular player

Note the two levels (this is why we call them hierarchical): 1) Player to player variability and 2) variability due to luck when batting. In a Bayesian framework, the first level is called a *prior distribution* and the second the *sampling distribution*.

Now, let's use this model for Jos's data. Suppose we want to predict his innate ability in the form of his *true* batting average θ. This would be the hierarchical model for our data:

$$\theta \sim N(.275, .027^2)$$

$$Y \mid \theta \sim N(\theta, .111^2)$$

We now are ready to compute a posterior distribution to summarize our prediction of θ. The continuous version of Bayes rule can be used here to derive the *posterior probability*, which is the distribution of the parameter θ given the observed data:

$$f_{\theta|Y}(\theta \mid Y) = \frac{f_{Y|\theta}(Y \mid \theta) f_\theta(\theta)}{f_Y(Y)}$$

$$= \frac{f_{Y|\theta}(Y \mid \theta) f_\theta(\theta)}{\int_\theta f_{Y|\theta}(Y \mid \theta) f_\theta(\theta)}$$

We are particularly interested in the θ that maximizes the posterior probability $f_{\theta|Y}(\theta \mid Y)$. In our case, we can show that the posterior is normal and we can compute the mean $E(\theta \mid y)$ and variance $\text{var}(\theta \mid y)$. Specifically, we can show the average of this distribution is the following:

$$E(\theta \mid y) = B\mu + (1 - B)Y$$

$$= \mu + (1 - B)(Y - \mu)$$

$$B = \frac{\sigma^2}{\sigma^2 + \tau^2}$$

It is a weighted average of the population average μ and the observed data Y. The weight depends on the SD of the population τ and the SD of our observed data σ. This weighted average is sometimes referred to as *shrinking* because it *shrinks* estimates towards a prior mean. In the case of Jos Iglesias, we have:

$$E(\theta \mid Y = .450) = B \times .275 + (1 - B) \times .450$$

$$= .275 + (1 - B)(.450 - .275)$$

$$B = \frac{.111^2}{.111^2 + .027^2} = 0.944$$

$$E(\theta \mid Y = 450) \approx .285$$

The variance can be shown to be:

$$\text{var}(\theta \mid y) = \frac{1}{1/\sigma^2 + 1/\tau^2} = \frac{1}{1/.111^2 + 1/.027^2} = 0.00069$$

and the standard deviation is therefore 0.026. So we started with a frequentist 95% confidence interval that ignored data from other players and summarized just Jos's data: .450 $\pm$ 0.220. Then we used a Bayesian approach that incorporated data from other players and other years to obtain a posterior probability. This is actually referred to as an empirical Bayes approach because we used data to construct the prior. From the posterior we can report what is called a 95% credible interval by reporting a region, centered at the mean, with a 95% chance of occurring. In our case, this turns out to be: .285 $\pm$ 0.052.

The Bayesian credible interval suggests that if another team is impressed by the .450 observation, we should consider trading Jos as we are predicting he will be just slightly above average. Interestingly, the Red Sox traded Jos to the Detroit Tigers in July. Here are the Jos Iglesias batting averages for the next five months.

Month	At Bat	Hits	AVG
April	20	9	.450
May	26	11	.423
June	86	34	.395
July	83	17	.205
August	85	25	.294
September	50	10	.200
Total w/o April	330	97	.293

Although both intervals included the final batting average, the Bayesian credible interval provided a much more precise prediction. In particular, it predicted that he would not be as good the remainder of the season. {pagebreak}

7.7 Exercises

1. A test for cystic fibrosis has an accuracy of 99%. Specifically, we mean that:

$$\text{Prob}(+|D) = 0.99, \text{Prob}(-|\text{no } D) = 0.99$$

The cystic fibrosis rate in the general population is 1 in 3,900, $\text{Prob}(D) = 0.00025$

If we select a random person and they test positive, what is probability that they have cystic fibrosis $\text{Prob}(D|+)$? Hint: use Bayes Rule.

$$\Pr(A|B) = \frac{\Pr(B|A)\Pr(A)}{\Pr(B)}$$

2. (Advanced) First download some baseball statistics.

```
tmpfile <- tempfile()
tmpdir <- tempdir()
download.file("http://seanlahman.com/files/database/lahman-csv_2014-02-14.zip",tmpfile)
##this shows us files
filenames <- unzip(tmpfile,list=TRUE)
players <- read.csv(unzip(tmpfile,files="Batting.csv",exdir=tmpdir),as.is=TRUE)
unlink(tmpdir)
file.remove(tmpfile)
```

We will use the **dplyr**, which you can read about here[7] to obtain data from 2010, 2011, and 2012, with more than 500 at bats (AB >= 500).

```
dat <- filter(players,yearID>=2010, yearID <=2012) %>% mutate(AVG=H/AB) %>% filter(AB>500)
```

What is the average of these batting averages?

3. What is the standard deviation of these batting averages?

4. Use exploratory data analysis to decide which of the following distributions approximates our AVG:

---

[7]http://cran.rstudio.com/web/packages/dplyr/vignettes/introduction.html

    (a) Normal.

    (b) Poisson.

    (c) F-distribution.

    (d) Uniform.

5. It is April and after 20 at bats, Jos Iglesias is batting .450 (which is very good). We can think of this as a binomial distribution with 20 trials, with probability of success p. Our sample estimate of p is .450. What is our estimate of standard deviation? Hint: This is the sum that is binomial divided by 20.

6. The Binomial is approximated by normal, so our sampling distribution is approximately normal with mean $Y = 0.45$ and SD $\sigma = 0.11$. Earlier we used a baseball database to determine that our prior distribution is Normal with mean $\mu = 0.275$ and SD $\tau = 0.027$. We also saw that this is the posterior mean prediction of the batting average.

What is your Bayes prediction for the batting average going forward?

$$E(\theta|y) = B\mu + (1 - B)Y$$
$$= \mu + (1 - B)(Y - \mu)$$
$$B = \frac{\sigma^2}{\sigma^2 + \tau^2}$$

7.8 Hierarchical Models

In this section, we use the mathematical theory which describes an approach that has become widely applied in the analysis of high-throughput data. The general idea is to build a *hierachichal model* with two levels. One level describes variability across samples/units, and the other describes variability across features. This is similar to the baseball example in which the first level described variability across players and the second described the randomness for the success of one player. The first level of variation is accounted for by all the models and approaches we have described here, for example the model that leads to the t-test. The second level provides power by permitting us to "borrow" information from all features to inform the inference performed on each one.

Here we describe one specific case that is currently the most widely used approach to inference with gene expression data. It is the model implemented by the `limma` Bioconductor package. This idea has been adapted to develop methods for other data types such as RNAseq by, for example, edgeR[8] and DESeq2[9]. This package provides an alternative to the t-test that greatly improves power by modeling the variance. While in the baseball example we modeled averages, here we model variances. Modelling variances requires more advanced math, but the concepts are practically the same. We motivate and demonstrate the approach with an example.

Here is a volcano showing effect sizes and p-value from applying a t-test to data from an experiment running six replicated samples with 16 genes artificially made to be different in two groups of three samples each. These 16 genes are the only genes for which the alternative hypothesis is true. In the plot they are shown in blue.

---

[8]http://www.ncbi.nlm.nih.gov/pubmed/19910308
[9]http://www.ncbi.nlm.nih.gov/pubmed/25516281

```
library(SpikeInSubset) ##Available from Bioconductor
data(rma95)
library(genefilter)
fac <- factor(rep(1:2,each=3))
tt <- rowttests(exprs(rma95),fac)
smallp <- with(tt, p.value < .01)
spike <- rownames(rma95) %in% colnames(pData(rma95))
cols <- ifelse(spike,"dodgerblue",ifelse(smallp,"red","black"))

with(tt, plot(-dm, -log10(p.value), cex=.8, pch=16,
    xlim=c(-1,1), ylim=c(0,4.5),
    xlab="difference in means",
    col=cols))
abline(h=2,v=c(-.2,.2), lty=2)
```

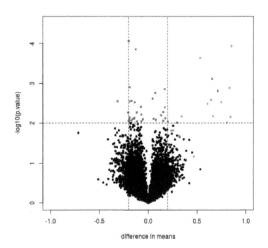

FIGURE 7.14
Volcano plot for t-test comparing two groups. Spiked-in genes are denoted with blue. Among the rest of the genes, those with p-value < 0.01 are denoted with red.

We cut-off the range of the y-axis at 4.5, but there is one blue point with a p-value smaller than 10^{-6}. Two findings stand out from this plot. The first is that only one of the positives would be found to be significant with a standard 5% FDR cutoff:

```
sum( p.adjust(tt$p.value,method = "BH")[spike] < 0.05)
```

```
## [1] 1
```

This of course has to do with the low power associated with a sample size of three in each group. The second finding is that if we forget about inference and simply rank the genes based on the size of the t-statistic, we obtain many false positives in any rank list of size larger than 1. For example, six of the top 10 genes ranked by t-statistic are false positives.

```
table( top50=rank(tt$p.value)<= 10, spike) #t-stat and p-val rank is the same
```

```
##          spike
## top50  FALSE  TRUE
##   FALSE 12604    12
##   TRUE      6     4
```

In the plot we notice that these are mostly genes for which the effect size is relatively small, implying that the estimated standard error is small. We can confirm this with a plot:

```
tt$s <- apply(exprs(rma95), 1, function(row)
  sqrt(.5 * (var(row[1:3]) + var(row[4:6]) ) ) )
with(tt, plot(s, -log10(p.value), cex=.8, pch=16,
              log="x",xlab="estimate of standard deviation",
              col=cols))
```

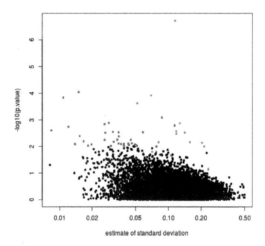

FIGURE 7.15
p-values versus standard deviation estimates.

Here is where a hierarchical model can be useful. If we can make an assumption about the distribution of these variances across genes, then we can improve estimates by "adjusting" estimates that are "too small" according to this distribution. In a previous section we described how the F-distribution approximates the distribution of the observed variances.

$$s^2 \sim s_0^2 F_{d,d_0}$$

Because we have thousands of data points, we can actually check this assumption and also estimate the parameters s_0 and d_0. This particular approach is referred to as empirical Bayes because it can be described as using data (empirical) to build the prior distribution (Bayesian approach).

Now we apply what we learned with the baseball example to the standard error estimates. As before we have an observed value for each gene s_g, a sampling distribution as a prior distribution. We can therefore compute a posterior distribution for the variance σ_g^2 and obtain the posterior mean. You can see the details of the derivation in this paper[10].

---

[10]http://www.ncbi.nlm.nih.gov/pubmed/16646809

$$\mathrm{E}[\sigma_g^2 \mid s_g] = \frac{d_0 s_0^2 + d s_g^2}{d_0 + d}$$

As in the baseball example, the posterior mean *shrinks* the observed variance s_g^2 towards the global variance s_0^2 and the weights depend on the sample size through the degrees of freedom d and, in this case, the shape of the prior distribution through d_0.

In the plot above we can see how the variance estimates *shrink* for 40 genes (code not shown):

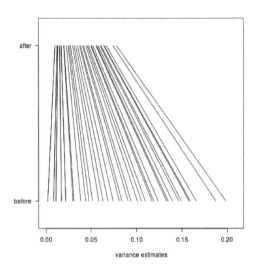

FIGURE 7.16
Illustration of how estimates shrink towards the prior expectation. Forty genes spanning the range of values were selected.

An important aspect of this adjustment is that genes having a sample standard deviation close to 0 are no longer close to 0 (they shrink towards s_0). We can now create a version of the t-test that instead of the sample standard deviation uses this posterior mean or "shrunken" estimate of the variance. We refer to these as *moderated* t-tests. Once we do this, the improvements can be seen clearly in the volcano plot:

```
library(limma)
fit <- lmFit(rma95, model.matrix(~ fac))
ebfit <- ebayes(fit)
limmares <- data.frame(dm=coef(fit)[,"fac2"], p.value=ebfit$p.value[,"fac2"])
with(limmares, plot(dm, -log10(p.value),cex=.8, pch=16,
    col=cols,xlab="difference in means",
    xlim=c(-1,1), ylim=c(0,5)))
abline(h=2,v=c(-.2,.2), lty=2)
```

The number of false positives in the top 10 is now reduced to 2.

```
table( top50=rank(limmares$p.value)<= 10, spike)
```

```
##          spike
## top50   FALSE  TRUE
##   FALSE 12608     8
##   TRUE      2     8
```

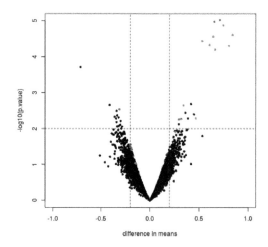

FIGURE 7.17
Volcano plot for moderated t-test comparing two groups. Spiked-in genes are denoted with blue. Among the rest of the genes, those with p-value < 0.01 are denoted with red.

7.9   Exercises

Load the following data (you can install it from Bioconductor) and extract the data matrix using `exprs`:

```
library(Biobase)
library(SpikeInSubset)
data(rma95)
y <- exprs(rma95)
```

This dataset comes from an experiment in which RNA was obtained from the same background pool to create six replicate samples. Then RNA from 16 genes were artificially added in different quantities to each sample. These quantities (in picoMolars) and gene IDs are stored here:

```
pData(rma95)
```

These quantities were the same in the first three arrays and in the last three arrays. So we define two groups like this:

```
g <- factor(rep(0:1,each=3))
```

and create an index of which rows are associated with the artificially added genes:

```
spike <- rownames(y) %in% colnames(pData(rma95))
```

1. Only these 16 genes are diferentially expressed since the six samples differ only due to sampling (they all come from the same background pool of RNA).

Perform a t-test on each gene using the `rowttest` function.

What proportion of genes with a p-value < 0.01 (no multiple comparison correction) are not part of the artificially added (false positive)?

2. Now compute the within group sample standard deviation for each gene (you can use group 1). Based on the p-value cut-off, split the genes into true positives, false positives, true negatives and false negatives. Create a boxplot comparing the sample SDs for each group. Which of the following best describes the boxplot?
   (a) The standard deviation is similar across all groups.
   (b) On average, the true negatives have much larger variability.
   (c) The false negatives have larger variability.
   (d) The false positives have smaller standard deviation.

3. In the previous two questions, we observed results consistent with the fact that the random variability associated with the sample standard deviation leads to t-statistics that are large by chance.

The sample standard deviation we use in the t-test is an estimate and with just a pair of triplicate samples, the variability associated with the denominator in the t-test can be large.

The following steps perform the basic `limma` analysis. We specify `coef=2` because we are interested in the difference between groups, not the intercept. The `eBayes` step uses a hierarchical model that provides a new estimate of the gene specific standard error.

```
library(limma)
fit <- lmFit(y, design=model.matrix(~ g))
colnames(coef(fit))
fit <- eBayes(fit)
```

Here is a plot of the original, new, hierarchical models based estimate versus the sample based estimate:

```
sampleSD = fit$sigma
posteriorSD = sqrt(fit$s2.post)
```

Which best describes what the hierarchical model does?

- A) Moves all the estimates of standard deviation closer to 0.12.

- B) Increases the estimates of standard deviation to increase t.

- C) Decreases the estimate of standard deviation.

- D) Decreases the effect size estimates.

4. Use these new estimates of standard deviation in the denominator of the t-test and compute p-values. You can do it like this:

```
library(limma)
fit = lmFit(y, design=model.matrix(~ g))
fit = eBayes(fit)
##second coefficient relates to diffences between group
pvals = fit$p.value[,2]
```

What proportion of genes with a p-value < 0.01 (no multiple comparison correction) are not part of the artificially added (false positive)?

Compare to the previous volcano plot and notice that we no longer have small p-values for genes with small effect sizes.

8

Distance and Dimension Reduction

8.1 Introduction

The concept of distance is quite intuitive. For example, when we cluster animals into sub-groups, we are implicitly defining a distance that permits us to say what animals are "close" to each other.

FIGURE 8.1
Clustering of animals.

Many of the analyses we perform with high-dimensional data relate directly or indirectly to distance. Many clustering and machine learning techniques rely on being able to define distance, using features or predictors. For example, to create *heatmaps*, which are widely used in genomics and other high-throughput fields, a distance is computed explicitly.

In these plots the measurements, which are stored in a matrix, are represented with colors after the columns and rows have been clustered. (A side note: red and green, a common color theme for heatmaps, are two of the most difficult colors for many color-blind people to discern.) Here we will learn the necessary mathematics and computing skills to understand and create heatmaps. We start by reviewing the mathematical definition of distance.

8.2 Euclidean Distance

As a review, let's define the distance between two points, A and B, on a Cartesian plane.

The euclidean distance between A and B is simply:

$$\sqrt{(A_x - B_x)^2 + (A_y - B_y)^2}$$

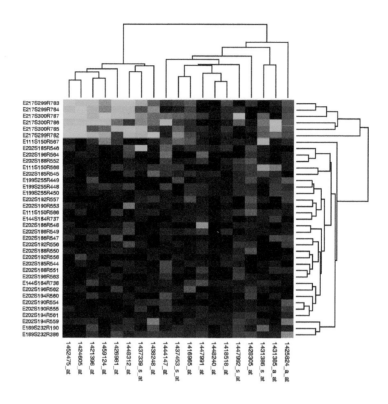

FIGURE 8.2
Example of heatmap. Image Source: Heatmap, Gaeddal, 01.28.2007 Wikimedia Commons

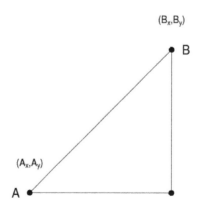

FIGURE 8.3

8.3   Distance in High Dimensions

We introduce a dataset with gene expression measurements for 22,215 genes from 189 samples. The R objects can be downloaded like this:

```
library(devtools)
install_github("genomicsclass/tissuesGeneExpression")
```

The data represent RNA expression levels for eight tissues, each with several individuals.

```
library(tissuesGeneExpression)
data(tissuesGeneExpression)
dim(e) ##e contains the expression data
```

```
## [1] 22215    189
```

```
table(tissue) ##tissue[i] tells us what tissue is represented by e[,i]
```

```
## tissue
##   cerebellum      colon endometrium hippocampus      kidney       liver
##           38         34          15          31          39          26
##     placenta
##            6
```

We are interested in describing distance between samples in the context of this dataset. We might also be interested in finding genes that *behave similarly* across samples.

To define distance, we need to know what the points are since mathematical distance is computed between points. With high dimensional data, points are no longer on the Cartesian plane. Instead they are in higher dimensions. For example, sample i is defined by a point in 22,215 dimensional space: $(Y_{1,i}, \ldots, Y_{22215,i})^\top$. Feature g is defined by a point in 189 dimensions $(Y_{g,1}, \ldots, Y_{g,189})^\top$

Once we define points, the Euclidean distance is defined in a very similar way as it is defined for two dimensions. For instance, the distance between two samples i and j is:

$$\text{dist}(i,j) = \sqrt{\sum_{g=1}^{22215} (Y_{g,i} - Y_{g,j})^2}$$

and the distance between two features h and g is:

$$\text{dist}(h,g) = \sqrt{\sum_{i=1}^{189} (Y_{h,i} - Y_{g,i})^2}$$

Note: In practice, distances between features are typically applied after standardizing the data for each feature. This is equivalent to computing one minus the correlation. This is done because the differences in overall levels between features are often not due to biological effects but technical ones. More details on this topic can be found in this presentation[1].

---

[1]http://master.bioconductor.org/help/course-materials/2002/Summer02Course/Distance/distance.pdf

Distance with matrix algebra The distance between samples i and j can be written as

$$\text{dist}(i, j) = (\mathbf{Y}_i - \mathbf{Y}_j)^\top (\mathbf{Y}_i - \mathbf{Y}_j)$$

with $\mathbf{Y}_i$ and $\mathbf{Y}_j$ columns i and j. This result can be very convenient in practice as computations can be made much faster using matrix multiplication.

Examples We can now use the formulas above to compute distance. Let's compute distance between samples 1 and 2, both kidneys, and then to sample 87, a colon.

```
x <- e[,1]
y <- e[,2]
z <- e[,87]
sqrt(sum((x-y)^2))
```

```
## [1] 85.8546
```

```
sqrt(sum((x-z)^2))
```

```
## [1] 122.8919
```

As expected, the kidneys are closer to each other. A faster way to compute this is using matrix algebra:

```
sqrt( crossprod(x-y) )
```

```
##             [,1]
## [1,] 85.8546
```

```
sqrt( crossprod(x-z) )
```

```
##             [,1]
## [1,] 122.8919
```

Now to compute all the distances at once, we have the function `dist`. Because it computes the distance between each row, and here we are interested in the distance between samples, we transpose the matrix

```
d <- dist(t(e))
class(d)
```

```
## [1] "dist"
```

Note that this produces an object of class `dist` and, to access the entries using row and column indices, we need to coerce it into a matrix:

```
as.matrix(d)[1,2]
```

```
## [1] 85.8546
```

```
as.matrix(d)[1,87]
```

```
## [1] 122.8919
```

It is important to remember that if we run `dist` on `e`, it will compute all pairwise distances between genes. This will try to create a 22215×22215 matrix that may crash your R sessions.

8.4 Exercises

If you have not done so already, install the data package `tissueGeneExpression`:

```
library(devtools)
install_github("genomicsclass/tissuesGeneExpression")
```

The data represents RNA expression levels for eight tissues, each with several *biological replictes*. We call samples that we consider to be from the same population, such as liver tissue from different individuals, *biological replicates*:

```
library(tissuesGeneExpression)
data(tissuesGeneExpression)
head(e)
head(tissue)
```

1. How many biological replicates for hippocampus?
2. What is the distance between samples 3 and 45?
3. What is the distance between gene 210486_at and 200805_at
4. If I run the command (don't run it!):

```
d = as.matrix( dist(e) )
```

how many cells (number of rows times number of columns) will this matrix have?

5. Compute the distance between all pair of samples:

```
d = dist( t(e) )
```

Read the help file for `dist`.
How many distances are stored in `d`? Hint: What is the length of d?

6. Why is the answer to exercise 5 not `ncol(e)^2`?
   (a) R made a mistake there.
   (b) Distances of 0 are left out.
   (c) Because we take advantage of symmetry: only the lower triangular matrix of the full distance matrix is stored thus, only `ncol(e)*(ncol(e)-1)/2` values.
   (d) Because it is equal`nrow(e)^2`

8.5 Dimension Reduction Motivation

Visualizing data is one of the most, if not the most, important step in the analysis of high-throughput data. The right visualization method may reveal problems with the experimental data that can render the results from a standard analysis, although typically appropriate, completely useless.

We have shown methods for visualizing global properties of the columns or rows, but plots that reveal relationships between columns or between rows are more complicated due to the high dimensionality of data. For example, to compare each of the 189 samples to each other, we would have to create, for example, 17,766 MA-plots. Creating one single scatterplot of the data is impossible since points are very high dimensional.

We will describe powerful techniques for exploratory data analysis based on *dimension reduction*. The general idea is to reduce the dataset to have fewer dimensions, yet approximately preserve important properties, such as the distance between samples. If we are able to reduce down to, say, two dimensions, we can then easily make plots. The technique behind it all, the singular value decomposition (SVD), is also useful in other contexts. Before introducing the rather complicated mathematics behind the SVD, we will motivate the ideas behind it with a simple example.

Example: Reducing two dimensions to one We consider an example with twin heights. Here we simulate 100 two dimensional points that represent the number of standard deviations each individual is from the mean height. Each point is a pair of twins:

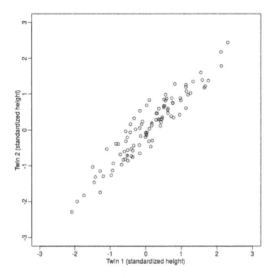

FIGURE 8.4
Simulated twin pair heights.

To help with the illustration, think of this as high-throughput gene expression data with the twin pairs representing the N samples and the two heights representing gene expression from two genes.

We are interested in the distance between any two samples. We can compute this using `dist`. For example, here is the distance between the two orange points in the figure above:

```
d=dist(t(y))
as.matrix(d)[1,2]
```

```
## [1] 1.140897
```

What if making two dimensional plots was too complex and we were only able to make 1 dimensional plots. Can we, for example, reduce the data to a one dimensional matrix that preserves distances between points?

If we look back at the plot, and visualize a line between any pair of points, the length of this line is the distance between the two points. These lines tend to go along the direction of the diagonal. We have seen before that we can "rotate" the plot so that the diagonal is in the x-axis by making a MA-plot instead:

```
z1 = (y[1,]+y[2,])/2 #the sum
z2 = (y[1,]-y[2,])    #the difference

z = rbind( z1, z2) #matrix now same dimensions as y

thelim <- c(-3,3)
mypar(1,2)

plot(y[1,],y[2,],xlab="Twin 1 (standardized height)",
    ylab="Twin 2 (standardized height)",
    xlim=thelim,ylim=thelim)
points(y[1,1:2],y[2,1:2],col=2,pch=16)

plot(z[1,],z[2,],xlim=thelim,ylim=thelim,xlab="Average height",ylab="Difference in height")
points(z[1,1:2],z[2,1:2],col=2,pch=16)
```

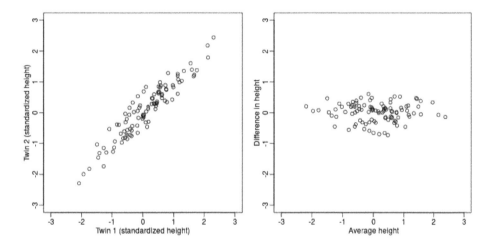

FIGURE 8.5
Twin height scatterplot (left) and MA-plot (right).

Later, we will start using linear algebra to represent transformation of the data such as this. Here we can see that to get **z** we multiplied **y** by the matrix:

$$A = \begin{pmatrix} 1/2 & 1/2 \\ 1 & -1 \end{pmatrix} \implies z = Ay$$

Remember that we can transform back by simply multiplying by A^{-1} as follows:

$$A^{-1} = \begin{pmatrix} 1 & 1/2 \\ 1 & -1/2 \end{pmatrix} \implies y = A^{-1}z$$

Rotations In the plot above, the distance between the two orange points remains roughly the same, relative to the distance between other points. This is true for all pairs of points. A simple re-scaling of the transformation we performed above will actually make the distances exactly the same. What we will do is multiply by a scalar so that the standard deviations of each point is preserved. If you think of the columns of **y** as independent random variables with standard deviation σ, then note that the standard deviations of M and A are:

$$\text{sd}[Z_1] = \text{sd}[(Y_1 + Y_2)/2] = \frac{1}{\sqrt{2}}\sigma \text{ and } \text{sd}[Z_2] = \text{sd}[Y_1 - Y_2] = \sqrt{2}\sigma$$

This implies that if we change the transformation above to:

$$A = \frac{1}{\sqrt{2}} \begin{pmatrix} 1 & 1 \\ 1 & -1 \end{pmatrix}$$

then the SD of the columns of Y are the same as the variance of the columns Z. Also, notice that $A^{-1}A = I$. We call matrices with these properties *orthogonal* and it guarantees the SD-preserving properties described above. The distances are now exactly preserved:

```
A <- 1/sqrt(2)*matrix(c(1,1,1,-1),2,2)
z <- A%*%y
d <- dist(t(y))
d2 <- dist(t(z))
mypar(1,1)
plot(as.numeric(d),as.numeric(d2)) #as.numeric turns distances into long vector
abline(0,1,col=2)
```

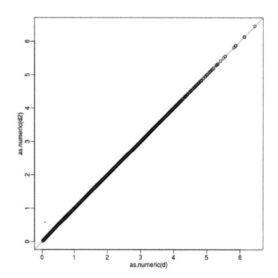

FIGURE 8.6
Distance computed from original data and after rotation is the same.

We call this particular transformation a *rotation* of y.

```
mypar(1,2)

thelim <- c(-3,3)
plot(y[1,],y[2,],xlab="Twin 1 (standardized height)",
    ylab="Twin 2 (standardized height)",
    xlim=thelim,ylim=thelim)
points(y[1,1:2],y[2,1:2],col=2,pch=16)

plot(z[1,],z[2,],xlim=thelim,ylim=thelim,xlab="Average height",ylab="Difference in height")
points(z[1,1:2],z[2,1:2],col=2,pch=16)
```

The reason we applied this transformation in the first place was because we noticed that to compute the distances between points, we followed a direction along the diagonal

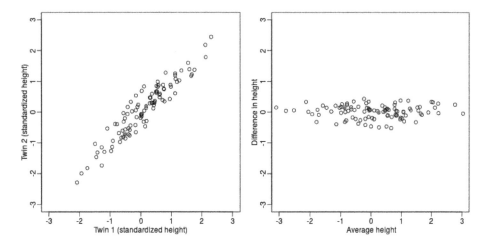

FIGURE 8.7
Twin height scatterplot (left) and after rotation (right).

in the original plot, which after the rotation falls on the horizontal, or the first dimension of **z**. So this rotation actually achieves what we originally wanted: we can preserve the distances between points with just one dimension. Let's remove the second dimension of **z** and recompute distances:

```
d3 = dist(z[1,]) ##distance computed using just first dimension
mypar(1,1)
plot(as.numeric(d),as.numeric(d3))
abline(0,1)
```

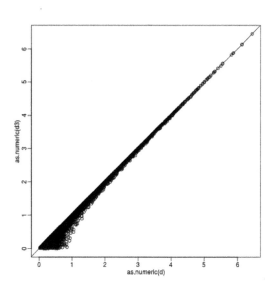

FIGURE 8.8
Distance computed with just one dimension after rotation versus actual distance.

The distance computed with just the one dimension provides a very good approximation to the actual distance and a very useful dimension reduction: from 2 dimensions to 1. This first dimension of the transformed data is actually the first *principal component*. This idea motivates the use of principal component analysis (PCA) and the singular value decomposition (SVD) to achieve dimension reduction more generally.

Important note on a difference to other explanations If you search the web for descriptions of PCA, you will notice a difference in notation to how we describe it here. This mainly stems from the fact that it is more common to have rows represent units. Hence, in the example shown here, Y would be transposed to be an $N \times 2$ matrix. In statistics this is also the most common way to represent the data: individuals in the rows. However, for practical reasons, in genomics it is more common to represent units in the columns. For example, genes are rows and samples are columns. For this reason, in this book we explain PCA and all the math that goes with it in a slightly different way than it is usually done. As a result, many of the explanations you find for PCA start out with the sample covariance matrix usually denoted with $\mathbf{X}^\top \mathbf{X}$ and having cells representing covariance between two units. Yet for this to be the case, we need the rows of $\mathbf{X}$ to represents units. So in our notation above, you would have to compute, after scaling, $\mathbf{YY}^\top$ instead.

Basically, if you want our explanations to match others you have to transpose the matrices we show here.

8.6 Singular Value Decomposition

In the previous section, we motivated dimension reduction and showed a transformation that permitted us to approximate the distance between two dimensional points with just one dimension. The singular value decomposition (SVD) is a generalization of the algorithm we used in the motivational section. As in the example, the SVD provides a transformation of the original data. This transformation has some very useful properties.

The main result SVD provides is that we can write an $m \times n$, matrix $\mathbf{Y}$ as

$$\mathbf{U}^\top \mathbf{Y} = \mathbf{D}\mathbf{V}^\top$$

With:

- $\mathbf{U}$ is an $m \times p$ orthogonal matrix
- $\mathbf{V}$ is an $p \times p$ orthogonal matrix
- $\mathbf{D}$ is an $n \times p$ diagonal matrix

with $p = \min(m, n)$. $\mathbf{U}^\top$ provides a rotation of our data $\mathbf{Y}$ that turns out to be very useful because the variability (sum of squares to be precise) of the columns of $\mathbf{U}^\top \mathbf{Y} = \mathbf{VD}$ are decreasing. Because $\mathbf{U}$ is orthogonal, we can write the SVD like this:

$$\mathbf{Y} = \mathbf{UDV}^\top$$

In fact, this formula is much more commonly used. We can also write the transformation like this:

$$\mathbf{YV} = \mathbf{UD}$$

This transformation of Y also results in a matrix with column of decreasing sum of squares.

Applying the SVD to the motivating example we have:

```
library(rafalib)
library(MASS)
n <- 100
y <- t(mvrnorm(n,c(0,0), matrix(c(1,0.95,0.95,1),2,2)))
s <- svd(y)
```

We can immediately see that applying the SVD results in a transformation very similar to the one we used in the motivating example:

```
round(sqrt(2) * s$u , 3)
```

```
##          [,1]    [,2]
## [1,] -0.994 -1.006
## [2,] -1.006  0.994
```

The plot we showed after the rotation was showing what we call the *principal components*: the second plotted against the first. To obtain the principal components from the SVD, we simply need the columns of the rotation $\mathbf{U}^\top \mathbf{Y}$:

```
PC1 = s$d[1]*s$v[,1]
PC2 = s$d[2]*s$v[,2]
plot(PC1,PC2,xlim=c(-3,3),ylim=c(-3,3))
```

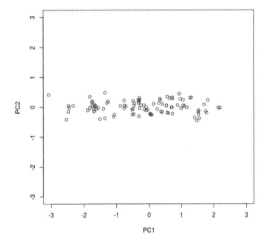

FIGURE 8.9
Second PC plotted against first PC for the twins height data.

How is this useful? It is not immediately obvious how incredibly useful the SVD can be, so let's consider some examples. In this example, we will greatly reduce the dimension of V and still be able to reconstruct Y.

Let's compute the SVD on the gene expression table we have been working with. We will take a subset of 100 genes so that computations are faster.

```
library(tissuesGeneExpression)
data(tissuesGeneExpression)
set.seed(1)
ind <- sample(nrow(e),500)
Y <- t(apply(e[ind,],1,scale)) #standardize data for illustration
```

The `svd` command returns the three matrices (only the diagonal entries are returned for D)

```
s <- svd(Y)
U <- s$u
V <- s$v
D <- diag(s$d) ##turn it into a matrix
```

First note that we can in fact reconstruct y:

```
Yhat <- U %*% D %*% t(V)
resid <- Y - Yhat
max(abs(resid))
```

```
## [1] 3.508305e-14
```

If we look at the sum of squares of **UD**, we see that the last few are quite close to 0 (perhaps we have some replicated columns).

```
plot(s$d)
```

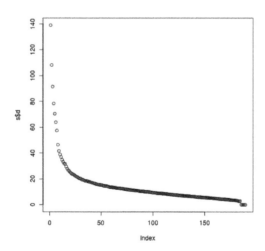

FIGURE 8.10
Entries of the diagonal of D for gene expression data.

This implies that the last columns of **V** have a very small effect on the reconstruction of **Y**. To see this, consider the extreme example in which the last entry of V is 0. In this case the last column of V is not needed at all. Because of the way the SVD is created, the columns of V have less and less influence on the reconstruction of Y. You commonly see this described as "explaining less variance". This implies that for a large matrix, by the time you get to the last columns, it is possible that there is not much left to "explain" As an example, we will look at what happens if we remove the four last columns:

```
k <- ncol(U)-4
Yhat <- U[,1:k] %*% D[1:k,1:k] %*% t(V[,1:k])
resid <- Y - Yhat
max(abs(resid))
```

```
## [1] 3.508305e-14
```

The largest residual is practically 0, meaning that we `Yhat` is practically the same as Y, yet we need 4 fewer dimensions to transmit the information.

By looking at d, we can see that, in this particular dataset, we can obtain a good approximation keeping only 94 columns. The following plots are useful for seeing how much of the variability is explained by each column:

```
plot(s$d^2/sum(s$d^2)*100,ylab="Percent variability explained")
```

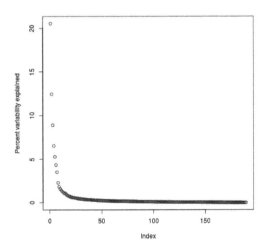

FIGURE 8.11
Percent variance explained by each principal component of gene expression data.

We can also make a cumulative plot:

```
plot(cumsum(s$d^2)/sum(s$d^2)*100,ylab="Percent variability explained",ylim=c(0,100),type="l")
```

Although we start with 189 dimensions, we can approximate Y with just 95:

```
k <- 95 ##out a possible 189
Yhat <- U[,1:k] %*% D[1:k,1:k] %*% t(V[,1:k])
resid <- Y - Yhat
boxplot(resid,ylim=quantile(Y,c(0.01,0.99)),range=0)
```

Therefore, by using only half as many dimensions, we retain most of the variability in our data:

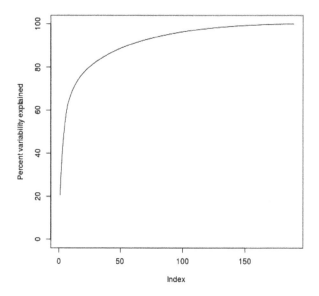

FIGURE 8.12
Cumulative variance explained by principal components of gene expression data.

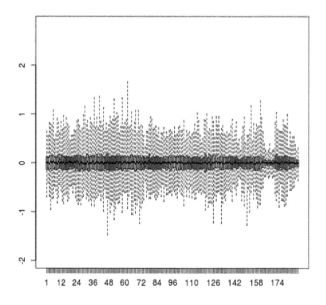

FIGURE 8.13
Residuals from comparing a reconstructed gene expression table using 95 PCs to the original data with 189 dimensions.

```
var(as.vector(resid))/var(as.vector(Y))
```

```
## [1] 0.04076899
```

We say that we explain 96% of the variability.

Note that we can compute this proportion from D:

```
1-sum(s$d[1:k]^2)/sum(s$d^2)
```

```
## [1] 0.04076899
```

The entries of D therefore tell us how much each PC contributes in term of variability explained.

Highly correlated data To help understand how the SVD works, we construct a dataset with two highly correlated columns.

For example:

```
m <- 100
n <- 2
x <- rnorm(m)
e <- rnorm(n*m,0,0.01)
Y <- cbind(x,x)+e
cor(Y)
```

```
##             x           x
## x 1.0000000 0.9998873
## x 0.9998873 1.0000000
```

In this case, the second column adds very little "information" since all the entries of `Y[,1]-Y[,2]` are close to 0. Reporting `rowMeans(Y)` is even more efficient since `Y[,1]-rowMeans(Y)` and `Y[,2]-rowMeans(Y)` are even closer to 0. `rowMeans(Y)` turns out to be the information represented in the first column on U. The SVD helps us notice that we explain almost all the variability with just this first column:

```
d <- svd(Y)$d
d[1]^2/sum(d^2)
```

```
## [1] 0.9999441
```

In cases with many correlated columns, we can achieve great dimension reduction:

```
m <- 100
n <- 25
x <- rnorm(m)
e <- rnorm(n*m,0,0.01)
Y <- replicate(n,x)+e
d <- svd(Y)$d
d[1]^2/sum(d^2)
```

```
## [1] 0.9999047
```

8.7 Exercises

For these exercises we are again going to use:

```
library(tissuesGeneExpression)
data(tissuesGeneExpression)
```

Before we start these exercises, it is important to reemphasize that when using the SVD, in practice the solution to SVD is not unique. This is because $\mathbf{U}\mathbf{D}\mathbf{V}^\top = (-\mathbf{U})\mathbf{D}(-\mathbf{V})^\top$. In fact, we can flip the sign of each column of $\mathbf{U}$ and, as long as we also flip the respective column in $\mathbf{V}$, we will arrive at the same solution. Here is an example:

```
s = svd(e)
signflips = sample(c(-1,1),ncol(e),replace=TRUE)
signflips
```

Now we switch the sign of each column and check that we get the same answer. We do this using the function sweep. If x is a matrix and a is a vector then sweep(x,1,a,FUN="*") applies the function FUN to each row i: FUN(x[i,],a[i]). In this case x[i,]*a[i]. If instead of 1 we use 2, sweep applies this to columns. To learn about sweep read ?sweep.

```
newu= sweep(s$u,2,signflips,FUN="*")
newv= sweep(s$v,2,signflips,FUN="*" )
identical( s$u %*% diag(s$d) %*% t(s$v), newu %*% diag(s$d) %*% t(newv))
```

This is important to know because different implementations of the SVD algorithm may give different signs, which can lead to the same code resulting in different answers when run in different computer systems.

1. Compute the SVD of e

```
s = svd(e)
```

Now compute the mean of each row:

```
m = rowMeans(e)
```

What is the correlation between the first column of $\mathbf{U}$ and m?

2. In exercise 1, we saw how the first column relates to the mean of the rows of e. If we change these means, the distances between columns do not change. For example, changing the means does not change the distances:

```
newmeans = rnorm(nrow(e)) ##random values we will add to create new means
newe = e+newmeans ##we change the means
sqrt(crossprod(e[,3]-e[,45]))
sqrt(crossprod(newe[,3]-newe[,45]))
```

So we might as well make the mean of each row 0, since it does not help us approximate the column distances. We will define y as the *detrended* e and recompute the SVD:

```
y = e - rowMeans(e)
s = svd(y)
```

We showed that $\mathbf{UDV}^\top$ is equal to y up to numerical error:

```
resid = y - s$u %*% diag(s$d) %*% t(s$v)
max(abs(resid))
```

The above can be made more efficient in two ways. First, using the `crossprod` and, second, not creating a diagonal matrix. In R, we can multiply matrices x by vector a. The result is a matrix with rows i equal to `x[i,]*a[i]`. Run the following example to see this.

```
x=matrix(rep(c(1,2),each=5),5,2)
x*c(1:5)
```

which is equivalent to:

```
sweep(x,1,1:5,"*")
```

This means that we don't have to convert s$d into a matrix.
Which of the following gives us the same as `diag(s$d)%*%t(s$v)` ?

- A) `s$d %*% t(s$v)`

- B) `s$d * t(s$v)`

- C) `t(s$d * s$v)`

- D) `s$v * s$d`

3. If we define `vd = t(s$d * t(s$v))`, then which of the following is not the same $UDV^\top$:

- A) `tcrossprod(s$u,vd)`

- B) `s$u %*% s$d * t(s$v)`

- C) `s$u %*% (s$d * t(s$v) )`

- D) `tcrossprod( t( s$d*t(s$u)) , s$v)`

4. Let `z = s$d * t(s$v)`. We showed a derivation demonstrating that because $\mathbf{U}$ is orthogonal, the distance between `e[,3]` and `e[,45]` is the same as the distance between `y[,3]` and `y[,45]`, which is the same as `vd[,3]` and `vd[,45]`

```
z = s$d * t(s$v)
##d was defined in question 2.1.5
sqrt(crossprod(e[,3]-e[,45]))
sqrt(crossprod(y[,3]-y[,45]))
sqrt(crossprod(z[,3]-z[,45]))
```

Note that the columns of **z** have 189 entries, compared to 22,215 for e. What is the difference, in absolute value, between the actual distance:

```
sqrt(crossprod(e[,3]-e[,45]))
```

and the approximation using only two dimensions of **z**?

5. How many dimensions do we need to use for the approximation in exercise 4 to be within 10%?

6. Compute distances between sample 3 and all other samples.

7. Recompute this distance using the two dimensional approximation. What is the Spearman correlation between this approximate distance and the actual distance?

The last exercise shows how just two dimensions can be useful to get a rough idea about the actual distances.

8.8 Projections

Now that we have described the concept of dimension reduction and some of the applications of SVD and principal component analysis, we focus on more details related to the mathematics behind these. We start with *projections*. A projection is a linear algebra concept that helps us understand many of the mathematical operations we perform on high-dimensional data. For more details, you can review projects in a linear algebra book. Here we provide a quick review and then provide some data analysis related examples.

As a review, remember that projections minimize the distance between points and subspace.

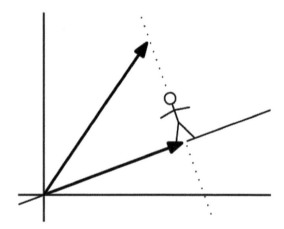

FIGURE 8.14
Illustration of projection.

We illustrate projections using a figure, in which the arrow on top is pointing to a point in space. In this particular cartoon, the space is two dimensional, but we should be thinking abstractly. The space is represented by the Cartesian plan and the line on which the little person stands is a subspace of points. The projection to this subspace is the place that is closest to the original point. Geometry tells us that we can find this closest point by dropping a perpendicular line (dotted line) from the point to the space. The little person is standing on the projection. The amount this person had to walk from the origin to the new projected point is referred to as *the coordinate*.

For the explanation of projections, we will use the standard matrix algebra notation for points: $\vec{y} \in \mathbb{R}^N$ is a point in N-dimensional space and $L \subset \mathbb{R}^N$ is smaller subspace.

Simple example with N=2 If we let $Y = \begin{pmatrix} 2 \\ 3 \end{pmatrix}$. We can plot it like this:

```
mypar (1,1)
plot(c(0,4),c(0,4),xlab="Dimension 1",ylab="Dimension 2",type="n")
arrows(0,0,2,3,lwd=3)
text(2,3," Y",pos=4,cex=3)
```

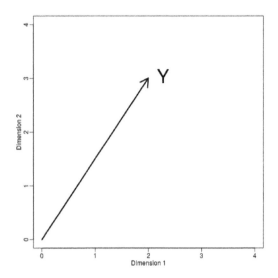

FIGURE 8.15
Geometric representation of Y.

We can immediately define a coordinate system by projecting this vector to the space defined by: $\begin{pmatrix} 1 \\ 0 \end{pmatrix}$ (the x-axis) and $\begin{pmatrix} 0 \\ 1 \end{pmatrix}$ (the y-axis). The projections of Y to the subspace defined by these points are 2 and 3 respectively:

$$Y = \begin{pmatrix} 2 \\ 3 \end{pmatrix}$$
$$= 2 \begin{pmatrix} 1 \\ 0 \end{pmatrix} + 3 \begin{pmatrix} 0 \\ 1 \end{pmatrix}$$

We say that 2 and 3 are the *coordinates* and that $\begin{pmatrix} 1 \\ 0 \end{pmatrix}$ and $\begin{pmatrix} 0 \\ 1 \end{pmatrix}$ are the bases.

Now let's define a new subspace. The red line in the plot below is subset L defined by points satisfying $c\vec{v}$ with $\vec{v} = \begin{pmatrix} 2 & 1 \end{pmatrix}^{\top}$. The projection of $\vec{y}$ onto L is the closest point on L to $\vec{y}$. So we need to find the c that minimizes the distance between $\vec{y}$ and $c\vec{v} = (2c, c)$. In linear algebra, we learn that the difference between these points is orthogonal to the space so:

$$(\vec{y} - \hat{c}\vec{v}) \cdot \vec{v} = 0$$

this implies that:

$$\vec{y} \cdot \vec{v} - \hat{c}\vec{v} \cdot \vec{v} = 0$$

and:

$$\hat{c} = \frac{\vec{y} \cdot \vec{v}}{\vec{v} \cdot \vec{v}}$$

Here the dot $\cdot$ represents the dot product: $\vec{x} \cdot \vec{y} = x_1 y_1 + \ldots x_n y_n$.

The following R code confirms this equation works:

```
mypar(1,1)
plot(c(0,4),c(0,4),xlab="Dimension 1",ylab="Dimension 2",type="n")
arrows(0,0,2,3,lwd=3)
abline(0,0.5,col="red",lwd=3) #if x=2c and y=c then slope is 0.5 (y=0.5x)
text(2,3," Y",pos=4,cex=3)
y=c(2,3)
x=c(2,1)
cc = crossprod(x,y)/crossprod(x)
segments(x[1]*cc,x[2]*cc,y[1],y[2],lty=2)
text(x[1]*cc,x[2]*cc,expression(hat(Y)),pos=4,cex=3)
```

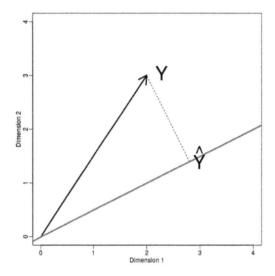

FIGURE 8.16
Projection of Y onto new subspace.

Note that if $\vec{v}$ was such that $\vec{v} \cdot \vec{v} = 1$, then $\hat{c}$ is simply $\vec{y} \cdot \vec{v}$ and the space L does not change. This simplification is one reason we like orthogonal matrices.

Example: The sample mean is a projection Let $\vec{y} \in \mathbb{R}^N$ and $L \subset \mathbb{R}^N$ is the space spanned by:

$$\vec{v} = \begin{pmatrix} 1 \\ \vdots \\ 1 \end{pmatrix} ; L = \{c\vec{v}; c \in \mathbb{R}\}$$

In this space, all components of the vectors are the same number, so we can think of this space as representing the constants: in the projection each dimension will be the same value. So what c minimizes the distance between $c\vec{v}$ and $\vec{y}$?

When talking about problems like this, we sometimes use two dimensional figures such as the one above. We simply abstract and think of $\vec{y}$ as a point in $N - dimensions$ and L as a subspace defined by a smaller number of values, in this case just one: c.

Getting back to our question, we know that the projection is:

$$\hat{c} = \frac{\vec{y} \cdot \vec{v}}{\vec{v} \cdot \vec{v}}$$

which in this case is the average:

$$\hat{c} = \frac{\vec{y} \cdot \vec{v}}{\vec{v} \cdot \vec{v}} = \frac{\sum_{i=1}^{N} Y_i}{\sum_{i=1}^{N} 1} = \bar{Y}$$

Here, it also would have been just as easy to use calculus:

$$\frac{\partial}{\partial c} \sum_{i=1}^{N} (Y_i - c)^2 = 0 \implies -2 \sum_{i=1}^{N} (Y_i - \hat{c}) = 0 \implies$$

$$Nc = \sum_{i=1}^{N} Y_i \implies \hat{c} = \bar{Y}$$

Example: Regression is also a projection Let us give a slightly more complicated example. Simple linear regression can also be explained with projections. Our data $\mathbf{Y}$ (we are no longer going to use the $\vec{y}$ notation) is again an N-dimensional vector and our model predicts Y_i with a line $\beta_0 + \beta_1 X_i$. We want to find the β_0 and β_1 that minimize the distance between Y and the space defined by:

$$L = \{\beta_0 \vec{v}_0 + \beta_1 \vec{v}_1; \vec{\beta} = (\beta_0, \beta_1) \in \mathbb{R}^2\}$$

with:

$$\vec{v}_0 = \begin{pmatrix} 1 \\ 1 \\ \vdots \\ 1 \end{pmatrix} \text{ and } \vec{v}_1 = \begin{pmatrix} X_1 \\ X_2 \\ \vdots \\ X_N \end{pmatrix}$$

Our $N \times 2$ matrix $\mathbf{X}$ is $[\vec{v}_0 \ \vec{v}_1]$ and any point in L can be written as $X\vec{\beta}$. The equation for the multidimensional version of orthogonal projection is:

$$X^\top (\vec{y} - X\vec{\beta}) = 0$$

which we have seen before and gives us:

$$X^\top X \hat{\beta} = X^\top \vec{y}$$

$$\hat{\beta} = (X^\top X)^{-1} X^\top \vec{y}$$

And the projection to L is therefore:

$$X(X^\top X)^{-1} X^\top \vec{y}$$

8.9 Rotations

One of the most useful applications of projections relates to coordinate rotations. In data analysis, simple rotations can result in easier to visualize and interpret data. We will describe the mathematics behind rotations and give some data analysis examples.

In our previous section, we used the following example:

$$Y = \begin{pmatrix} 2 \\ 3 \end{pmatrix} = 2 \begin{pmatrix} 1 \\ 0 \end{pmatrix} + 3 \begin{pmatrix} 0 \\ 1 \end{pmatrix}$$

and noted that 2 and 3 are the *coordinates*.

```
library(rafalib)
mypar()
plot(c(-2,4),c(-2,4),xlab="Dimension 1",ylab="Dimension 2",
     type="n",xaxt="n",yaxt="n",bty="n")
text(rep(0,6),c(c(-2,-1),c(1:4)),as.character(c(c(-2,-1),c(1:4))),pos=2)
text(c(c(-2,-1),c(1:4)),rep(0,6),as.character(c(c(-2,-1),c(1:4))),pos=1)
abline(v=0,h=0)
arrows(0,0,2,3,lwd=3)
segments(2,0,2,3,lty=2)
segments(0,3,2,3,lty=2)
text(2,3," Y",pos=4,cex=3)
```

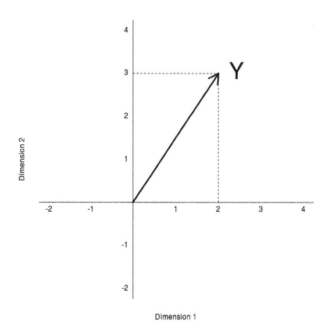

FIGURE 8.17
Plot of (2,3) as coordinates along Dimension 1 (1,0) and Dimension 2 (0,1).

However, mathematically we can represent the point $(2,3)$ with other linear combinations:

$$Y = \begin{pmatrix} 2 \\ 3 \end{pmatrix}$$

$$= 2.5 \begin{pmatrix} 1 \\ 1 \end{pmatrix} + -1 \begin{pmatrix} 0.5 \\ -0.5 \end{pmatrix}$$

The new coordinates are:

$$Z = \begin{pmatrix} 2.5 \\ -1 \end{pmatrix}$$

Graphically, we can see that the coordinates are the projections to the spaces defined by the new basis:

```
library(rafalib)
mypar()
plot(c(-2,4),c(-2,4),xlab="Dimension 1",ylab="Dimension 2",
    type="n",xaxt="n",yaxt="n",bty="n")
text(rep(0,6),c(c(-2,-1),c(1:4)),as.character(c(c(-2,-1),c(1:4))),pos=2)
text(c(c(-2,-1),c(1:4)),rep(0,6),as.character(c(c(-2,-1),c(1:4))),pos=1)
abline(v=0,h=0)
abline(0,1,col="red")
abline(0,-1,col="red")
arrows(0,0,2,3,lwd=3)
y=c(2,3)
x1=c(1,1)##new basis
x2=c(0.5,-0.5)##new basis
c1 = crossprod(x1,y)/crossprod(x1)
c2 = crossprod(x2,y)/crossprod(x2)
segments(x1[1]*c1,x1[2]*c1,y[1],y[2],lty=2)
segments(x2[1]*c2,x2[2]*c2,y[1],y[2],lty=2)
text(2,3," Y",pos=4,cex=3)
```

We can go back and forth between these two representations of $(2,3)$ using matrix multiplication.

$$Y = AZ$$

$$A^{-1}Y = Z$$

$$A = \begin{pmatrix} 1 & 0.5 \\ 1 & -0.5 \end{pmatrix} \implies A^{-1} = \begin{pmatrix} 0.5 & 0.5 \\ 1 & -1 \end{pmatrix}$$

Z and Y carry the same information, but in a different *coordinate system*.

Example: Twin heights Here are 100 two dimensional points Y

Here are the rotations: $Z = A^{-1}Y$

What we have done here is rotate the data so that the first coordinate of Z is the average height, while the second is the difference between twin heights.

We have used the singular value decomposition to find principal components. It is sometimes useful to think of the SVD as a rotation, for example $\mathbf{U}^\top\mathbf{Y}$, that gives us a new coordinate system $\mathbf{DV}^\top$ in which the dimensions are ordered by how much variance they explain.

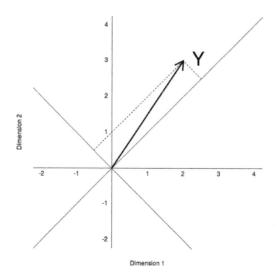

FIGURE 8.18
Plot of (2,3) as a vector in a rotatated space, relative to the original dimensions.

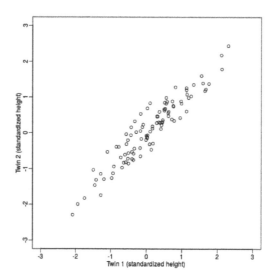

FIGURE 8.19
Twin 2 heights versus twin 1 heights.

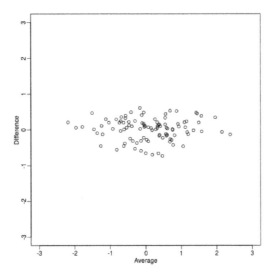

FIGURE 8.20
Rotation of twin 2 heights versus twin 1 heights.

8.10   Multi-Dimensional Scaling Plots

We will motivate multi-dimensional scaling (MDS) plots with a gene expression example. To simplify the illustration we will only consider three tissues:

```
library(rafalib)
library(tissuesGeneExpression)
data(tissuesGeneExpression)
colind <- tissue%in%c("kidney","colon","liver")
mat <- e[,colind]
group <- factor(tissue[colind])
dim(mat)
```

```
## [1] 22215    99
```

As an exploratory step, we wish to know if gene expression profiles stored in the columns of `mat` show more similarity between tissues than across tissues. Unfortunately, as mentioned above, we can't plot multi-dimensional points. In general, we prefer two-dimensional plots, but making plots for every pair of genes or every pair of samples is not practical. MDS plots become a powerful tool in this situation.

The math behind MDS Now that we know about SVD and matrix algebra, understanding MDS is relatively straightforward. For illustrative purposes let's consider the SVD decomposition:

$$\mathbf{Y} = \mathbf{UDV}^\top$$

and assume that the sum of squares of the first two columns $\mathbf{U}^\top\mathbf{Y} = \mathbf{DV}^\top$ is much larger than the sum of squares of all other columns. This can be written as: $d_1 + d_2 \gg d_3 + \cdots + d_n$ with d_i the i-th entry of the $\mathbf{D}$ matrix. When this happens, we then have:

$$\mathbf{Y} \approx [\mathbf{U}_1\mathbf{U}_2] \begin{pmatrix} d_1 & 0 \\ 0 & d_2 \end{pmatrix} [\mathbf{V}_1\mathbf{V}_2]^\top$$

This implies that column i is approximately:

$$\mathbf{Y}_i \approx [\mathbf{U}_1\mathbf{U}_2] \begin{pmatrix} d_1 & 0 \\ 0 & d_2 \end{pmatrix} \begin{pmatrix} v_{i,1} \\ v_{i,2} \end{pmatrix} = [\mathbf{U}_1\mathbf{U}_2] \begin{pmatrix} d_1 v_{i,1} \\ d_2 v_{i,2} \end{pmatrix}$$

If we define the following two dimensional vector...

$$\mathbf{Z}_i = \begin{pmatrix} d_1 v_{i,1} \\ d_2 v_{i,2} \end{pmatrix}$$

... then

$$
\begin{aligned}
(\mathbf{Y}_i - \mathbf{Y}_j)^\top (\mathbf{Y}_i - \mathbf{Y}_j) &\approx \{[\mathbf{U}_1\mathbf{U}_2](\mathbf{Z}_i - \mathbf{Z}_j)\}^\top \{[\mathbf{U}_1\mathbf{U}_2](\mathbf{Z}_i - \mathbf{Z}_j)\} \\
&= (\mathbf{Z}_i - \mathbf{Z}_j)^\top [\mathbf{U}_1\mathbf{U}_2]^\top [\mathbf{U}_1\mathbf{U}_2](\mathbf{Z}_i - \mathbf{Z}_j) \\
&= (\mathbf{Z}_i - \mathbf{Z}_j)^\top (\mathbf{Z}_i - \mathbf{Z}_j) \\
&= (Z_{i,1} - Z_{j,1})^2 + (Z_{i,2} - Z_{j,2})^2
\end{aligned}
$$

This derivation tells us that the distance between samples i and j is approximated by the distance between two dimensional points.

$$(\mathbf{Y}_i - \mathbf{Y}_j)^\top (\mathbf{Y}_i - \mathbf{Y}_j) \approx (Z_{i,1} - Z_{j,1})^2 + (Z_{i,2} - Z_{j,2})^2$$

Because Z is a two dimensional vector, we can visualize the distances between each sample by plotting $\mathbf{Z}_1$ versus $\mathbf{Z}_2$ and visually inspect the distance between points. Here is this plot for our example dataset:

```
s <- svd(mat-rowMeans(mat))
PC1 <- s$d[1]*s$v[,1]
PC2 <- s$d[2]*s$v[,2]
mypar(1,1)
plot(PC1,PC2,pch=21,bg=as.numeric(group))
legend("bottomright",levels(group),col=seq(along=levels(group)),pch=15,cex=1.5)
```

Note that the points separate by tissue type as expected. Now the accuracy of the approximation above depends on the proportion of variance explained by the first two principal components. As we showed above, we can quickly see this by plotting the variance explained plot:

```
plot(s$d^2/sum(s$d^2))
```

Although the first two PCs explain over 50% of the variability, there is plenty of information that this plot does not show. However, it is an incredibly useful plot for obtaining, via visualization, a general idea of the distance between points. Also, notice that we can plot other dimensions as well to search for patterns. Here are the 3rd and 4th PCs:

```
PC3 <- s$d[3]*s$v[,3]
PC4 <- s$d[4]*s$v[,4]
mypar(1,1)
plot(PC3,PC4,pch=21,bg=as.numeric(group))
legend("bottomright",levels(group),col=seq(along=levels(group)),pch=15,cex=1.5)
```

Note that the 4th PC shows a strong separation within the kidney samples. Later we will learn about batch effects, which might explain this finding.

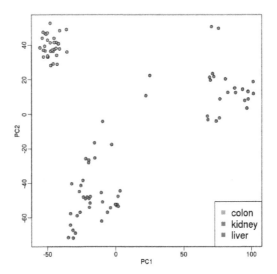

FIGURE 8.21
Multi-dimensional scaling (MDS) plot for tissue gene expression data.

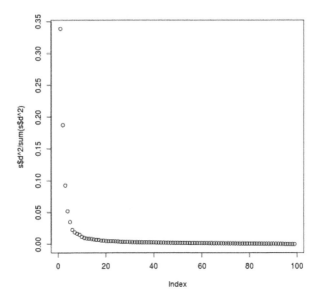

FIGURE 8.22
Variance explained for each principal component.

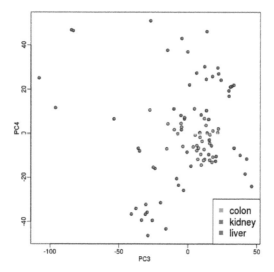

FIGURE 8.23
Third and fourth principal components.

`cmdscale` Although we used the `svd` functions above, there is a special function that is specifically made for MDS plots. It takes a distance object as an argument and then uses principal component analysis to provide the best approximation to this distance that can be obtained with k dimensions. This function is more efficient because one does not have to perform the full SVD, which can be time consuming. By default it returns two dimensions, but we can change that through the parameter `k` which defaults to 2.

```
d <- dist(t(mat))
mds <- cmdscale(d)
mypar()
plot(mds[,1],mds[,2],bg=as.numeric(group),pch=21,
    xlab="First dimension",ylab="Second dimension")
legend("bottomleft",levels(group),col=seq(along=levels(group)),pch=15)
```

These two approaches are equivalent up to an arbitrary sign change.

```
mypar(1,2)
for(i in 1:2){
  plot(mds[,i],s$d[i]*s$v[,i],main=paste("PC",i))
  b = ifelse( cor(mds[,i],s$v[,i]) > 0, 1, -1)
  abline(0,b) ##b is 1 or -1 depending on the arbitrary sign "flip"
}
```

Why the arbitrary sign? The SVD is not unique because we can multiply any column of $\mathbf{V}$ by -1 as long as we multiply the sample column of $\mathbf{U}$ by -1. We can see this immediately by noting that:

$$-1\mathbf{U}\mathbf{D}(-1)\mathbf{V}^{\top} = \mathbf{U}\mathbf{D}\mathbf{V}^{\top}$$

Why we substract the mean In all calculations above we subtract the row means before we compute the singular value decomposition. If what we are trying to do is approximate

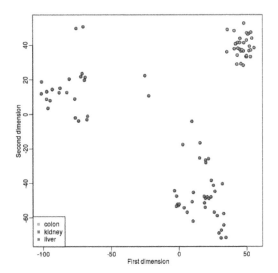

FIGURE 8.24
MDS computed with cmdscale function.

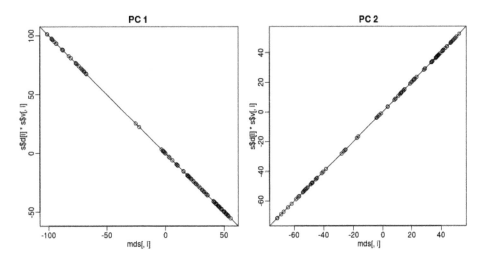

FIGURE 8.25
Comparison of MDS first two PCs to SVD first two PCs.

the distance between columns, the distance between $\mathbf{Y}_i$ and $\mathbf{Y}_j$ is the same as the distance between $\mathbf{Y}_i - \mu$ and $\mathbf{Y}_j - \mu$ since the μ cancels out when computing said distance:

$$\{(\mathbf{Y}_i - \mu) - (\mathbf{Y}_j - \mu)\}^\top \{(\mathbf{Y}_i - \mu) - (\mathbf{Y}_j - \mu)\} = \{\mathbf{Y}_i - \mathbf{Y}_j\}^\top \{\mathbf{Y}_i - \mathbf{Y}_j\}$$

Because removing the row averages reduces the total variation, it can only make the SVD approximation better. {pagebreak}

8.11 Exercises

1. Using the **z** we computed in exercise 4 of the previous exercises:

```
library(tissuesGeneExpression)
data(tissuesGeneExpression)
y = e - rowMeans(e)
s = svd(y)
z = s$d * t(s$v)
```

we can make an mds plot:

```
library(rafalib)
ftissue = factor(tissue)
mypar(1,1)
plot(z[1,],z[2,],col=as.numeric(ftissue))
legend("topleft",levels(ftissue),col=seq_along(ftissue),pch=1)
```

Now run the function `cmdscale` on the original data:

```
d = dist(t(e))
mds = cmdscale(d)
```

What is the absolute value of the correlation between the first dimension of **z** and the first dimension in mds?

2. What is the absolute value of the correlation between the second dimension of **z** and the second dimension in mds?

3. Load the following dataset:

```
library(GSE5859Subset)
data(GSE5859Subset)
```

Compute the svd and compute z.

```
s = svd(geneExpression-rowMeans(geneExpression))
z = s$d * t(s$v)
```

Which dimension of **z** most correlates with the outcome `sampleInfo$group`?

4. What is this max correlation?

5. Which dimension of **z** has the second highest correlation with the outcome `sampleInfo$group`?

6. Note these measurements were made during two months:

```
sampleInfo$date
```

We can extract the month this way:

```
month = format( sampleInfo$date, "%m")
month = factor( month)
```

Which dimension of **z** has the second highest correlation with the outcome `month`?

7. What is this correlation?

8. (Advanced) The same dimension is correlated with both the group and the date. The following are also correlated:

```
table(sampleInfo$g,month)
```

So is this first dimension related directly to group or is it related only through the month? Note that the correlation with month is higher. This is related to *batch effects* which we will learn about later.

In exercise 3 we saw that one of the dimensions was highly correlated to the `sampleInfo$group`. Now take the 5th column of **U** and stratify by the gene chromosome. Remove `chrUn` and make a boxplot of the values of $\mathbf{U}_5$ stratified by chromosome.

Which chromosome looks different from the rest? Copy and paste the name as it appears in `geneAnnotation`.

Given the answer to the last exercise, any guesses as to what `sampleInfo$group` represents?

8.12 Principal Component Analysis

We have already mentioned principal component analysis (PCA) above and noted its relation to the SVD. Here we provide further mathematical details.

Example: Twin heights We started the motivation for dimension reduction with a simulated example and showed a rotation that is very much related to PCA.

Here we explain specifically what are the principal components (PCs).

Let **Y** be $2 \times N$ matrix representing our data. The analogy is that we measure expression from 2 genes and each column is a sample. Suppose we are given the task of finding a 2×1 vector $\mathbf{u}_1$ such that $\mathbf{u}_1^\top \mathbf{v}_1 = 1$ and it maximizes $(\mathbf{u}_1^\top \mathbf{Y})^\top (\mathbf{u}_1^\top \mathbf{Y})$. This can be viewed as a projection of each sample or column of **Y** into the subspace spanned by $\mathbf{u}_1$. So we are looking for a transformation in which the coordinates show high variability.

Let's try $\mathbf{u} = (1,0)^\top$. This projection simply gives us the height of twin 1 shown in orange below. The sum of squares is shown in the title.

```
mypar(1,1)
plot(t(Y), xlim=thelim, ylim=thelim,
     main=paste("Sum of squares :",round(crossprod(Y[1,]),1)))
abline(h=0)
apply(Y,2,function(y) segments(y[1],0,y[1],y[2],lty=2))
```

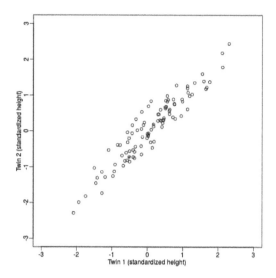

FIGURE 8.26
Twin heights scatter plot.

```
## NULL

points(Y[1,],rep(0,ncol(Y)),col=2,pch=16,cex=0.75)
```

Can we find a direction with higher variability? How about:
$\mathbf{u} = \begin{pmatrix} 1 \\ -1 \end{pmatrix}$? This does not satisfy $\mathbf{u}^\top \mathbf{u} = 1$ so let's instead try $\mathbf{u} = \begin{pmatrix} 1/\sqrt{2} \\ -1/\sqrt{2} \end{pmatrix}$

```
u <- matrix(c(1,-1)/sqrt(2),ncol=1)
w=t(u)%*%Y
mypar(1,1)
plot(t(Y),
     main=paste("Sum of squares:",round(tcrossprod(w),1)),xlim=thelim,ylim=thelim)
abline(h=0,lty=2)
abline(v=0,lty=2)
abline(0,-1,col=2)
Z = u%*%w
for(i in seq(along=w))
  segments(Z[1,i],Z[2,i],Y[1,i],Y[2,i],lty=2)
points(t(Z), col=2, pch=16, cex=0.5)
```

This relates to the difference between twins, which we know is small. The sum of squares confirms this.

Finally, let's try:
$\mathbf{u} = \begin{pmatrix} 1/\sqrt{2} \\ 1/\sqrt{2} \end{pmatrix}$

```
u <- matrix(c(1,1)/sqrt(2),ncol=1)
w=t(u)%*%Y
mypar()
plot(t(Y), main=paste("Sum of squares:",round(tcrossprod(w),1)),
     xlim=thelim, ylim=thelim)
```

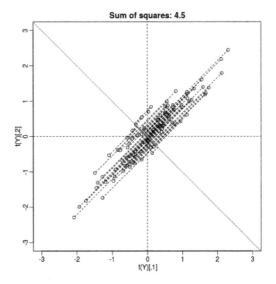

FIGURE 8.27
Data projected onto space spanned by (1 0).

```
abline(h=0,lty=2)
abline(v=0,lty=2)
abline(0,1,col=2)
points(u%*%w, col=2, pch=16, cex=1)
Z = u%*%w
for(i in seq(along=w))
  segments(Z[1,i], Z[2,i], Y[1,i], Y[2,i], lty=2)
points(t(Z),col=2,pch=16,cex=0.5)
```

This is a re-scaled average height, which has higher sum of squares. There is a mathematical procedure for determining which $\mathbf{v}$ maximizes the sum of squares and the SVD provides it for us.

The principal components The orthogonal vector that maximizes the sum of squares:

$$(\mathbf{u}_1^\top \mathbf{Y})^\top (\mathbf{u}_1^\top \mathbf{Y})$$

$\mathbf{u}_1^\top \mathbf{Y}$ is referred to as the first PC. The *weights* $\mathbf{u}$ used to obtain this PC are referred to as the *loadings*. Using the language of rotations, it is also referred to as the *direction* of the first PC, which are the new coordinates.

To obtain the second PC, we repeat the exercise above, but for the residuals:

$$\mathbf{r} = \mathbf{Y} - \mathbf{u}_1^\top \mathbf{Y} \mathbf{v}_1$$

The second PC is the vector with the following properties:

$$\mathbf{v}_2^\top \mathbf{v}_2 = 1$$

$$\mathbf{v}_2^\top \mathbf{v}_1 = 0$$

and maximizes $(\mathbf{r}\mathbf{v}_2)^\top \mathbf{r}\mathbf{v}_2$.

When Y is $N \times m$ we repeat to find 3rd, 4th, ..., m-th PCs.

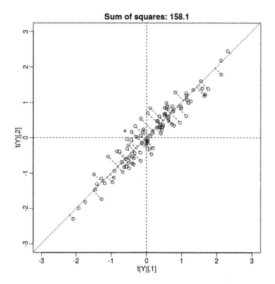

FIGURE 8.28
Data projected onto space spanned by first PC.

`prcomp` We have shown how to obtain PCs using the SVD. However, R has a function specifically designed to find the principal components. In this case, the data is centered by default. The following function:

```
pc <- prcomp( t(Y) )
```

produces the same results as the SVD up to arbitrary sign flips:

```
s <- svd( Y - rowMeans(Y) )
mypar(1,2)
for(i in 1:nrow(Y) ){
  plot(pc$x[,i], s$d[i]*s$v[,i])
}
```

The loadings can be found this way:

```
pc$rotation
```

```
##              PC1        PC2
## [1,] 0.7072304  0.7069831
## [2,] 0.7069831 -0.7072304
```

which are equivalent (up to a sign flip) to:

```
s$u
```

```
##            [,1]       [,2]
## [1,] -0.7072304 -0.7069831
## [2,] -0.7069831  0.7072304
```

The equivalent of the variance explained is included in the:

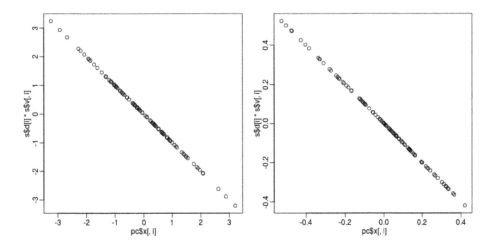

FIGURE 8.29
Plot showing SVD and prcomp give same results.

```
pc$sdev
```

```
## [1] 1.2542672 0.2141882
```

component.

We take the transpose of Y because `prcomp` assumes the previously discussed ordering: units/samples in row and features in columns.

9

Basic Machine Learning

Machine learning is a very broad topic and a highly active research area. In the life sciences, much of what is described as "precision medicine" is an application of machine learning to biomedical data. The general idea is to predict or discover outcomes from measured predictors. Can we discover new types of cancer from gene expression profiles? Can we predict drug response from a series of genotypes? Here we give a very brief introduction to two major machine learning components: clustering and class prediction. There are many good resources to learn more about machine learning, for example the excellent textbook *The Elements of Statistical Learning: Data Mining, Inference, and Prediction*, by Trevor Hastie, Robert Tibshirani and Jerome Friedman. A free PDF of this book can be found here[1].

9.1 Clustering

We will demonstrate the concepts and code needed to perform clustering analysis with the tissue gene expression data:

```
library(tissuesGeneExpression)
data(tissuesGeneExpression)
```

To illustrate the main application of clustering in the life sciences, let's pretend that we don't know these are different tissues and are interested in clustering. The first step is to compute the distance between each sample:

```
d <- dist( t(e) )
```

Hierarchical clustering With the distance between each pair of samples computed, we need clustering algorithms to join them into groups. Hierarchical clustering is one of the many clustering algorithms available to do this. Each sample is assigned to its own group and then the algorithm continues iteratively, joining the two most similar clusters at each step, and continuing until there is just one group. While we have defined distances between samples, we have not yet defined distances between groups. There are various ways this can be done and they all rely on the individual pairwise distances. The helpfile for `hclust` includes detailed information.

We can perform hierarchical clustering based on the distances defined above using the `hclust` function. This function returns an `hclust` object that describes the groupings that were created using the algorithm described above. The `plot` method represents these relationships with a tree or dendrogram:

---

[1]http://statweb.stanford.edu/~tibs/ElemStatLearn/

```
library(rafalib)
mypar()
hc <- hclust(d)
hc
```

```
##
## Call:
## hclust(d = d)
##
## Cluster method   : complete
## Distance         : euclidean
## Number of objects: 189
```

```
plot(hc,labels=tissue,cex=0.5)
```

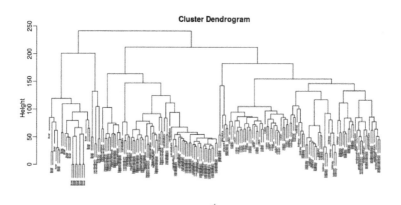

FIGURE 9.1
Dendrogram showing hierarchical clustering of tissue gene expression data.

Does this technique "discover" the clusters defined by the different tissues? In this plot, it is not easy to see the different tissues so we add colors by using the `myplclust` function from the `rafalib` package.

```
myplclust(hc, labels=tissue, lab.col=as.fumeric(tissue), cex=0.5)
```

Visually, it does seem as if the clustering technique has discovered the tissues. However, hierarchical clustering does not define specific clusters, but rather defines the dendrogram above. From the dendrogram we can decipher the distance between any two groups by looking at the height at which the two groups split into two. To define clusters, we need to "cut the tree" at some distance and group all samples that are within that distance into groups below. To visualize this, we draw a horizontal line at the height we wish to cut and this defines that line. We use 120 as an example:

```
myplclust(hc, labels=tissue, lab.col=as.fumeric(tissue),cex=0.5)
abline(h=120)
```

If we use the line above to cut the tree into clusters, we can examine how the clusters overlap with the actual tissues:

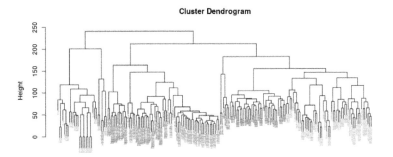

FIGURE 9.2
Dendrogram showing hierarchical clustering of tissue gene expression data with colors denoting tissues.

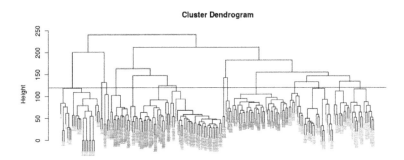

FIGURE 9.3
Dendrogram showing hierarchical clustering of tissue gene expression data with colors denoting tissues. Horizontal line defines actual clusters.

```
hclusters <- cutree(hc, h=120)
table(true=tissue, cluster=hclusters)
```

```
##                 cluster
## true             1  2  3  4  5  6  7  8  9 10 11 12 13 14
##    cerebellum    0  0  0  0 31  0  0  0  2  0  0  5  0  0
##    colon         0  0  0  0  0  0 34  0  0  0  0  0  0  0
##    endometrium   0  0  0  0  0  0  0  0  0  0 15  0  0  0
##    hippocampus   0  0 12 19  0  0  0  0  0  0  0  0  0  0
##    kidney        9 18  0  0  0 10  0  0  2  0  0  0  0  0
##    liver         0  0  0  0  0  0  0 24  0  2  0  0  0  0
##    placenta      0  0  0  0  0  0  0  0  0  0  0  0  2  4
```

We can also ask **cutree** to give us back a given number of clusters. The function then automatically finds the height that results in the requested number of clusters:

```
hclusters <- cutree(hc, k=8)
table(true=tissue, cluster=hclusters)
```

```
##                 cluster
## true             1  2  3  4  5  6  7  8
##    cerebellum    0  0 31  0  0  2  5  0
##    colon         0  0  0 34  0  0  0  0
##    endometrium  15  0  0  0  0  0  0  0
##    hippocampus   0 12 19  0  0  0  0  0
##    kidney       37  0  0  0  0  2  0  0
##    liver         0  0  0  0 24  2  0  0
##    placenta      0  0  0  0  0  0  0  6
```

In both cases we do see that, with some exceptions, each tissue is uniquely represented by one of the clusters. In some instances, the one tissue is spread across two tissues, which is due to selecting too many clusters. Selecting the number of clusters is generally a challenging step in practice and an active area of research.

K-means We can also cluster with the **kmeans** function to perform k-means clustering. As an example, let's run k-means on the samples in the space of the first two genes:

```
set.seed(1)
km <- kmeans(t(e[1:2,]), centers=7)
names(km)
```

```
## [1] "cluster"     "centers"    "totss"      "withinss"
## [5] "tot.withinss" "betweenss"  "size"       "iter"
## [9] "ifault"
```

```
mypar(1,2)
plot(e[1,], e[2,], col=as.fumeric(tissue), pch=16)
plot(e[1,], e[2,], col=km$cluster, pch=16)
```

In the first plot, color represents the actual tissues, while in the second, color represents the clusters that were defined by **kmeans**. We can see from tabulating the results that this particular clustering exercise did not perform well:

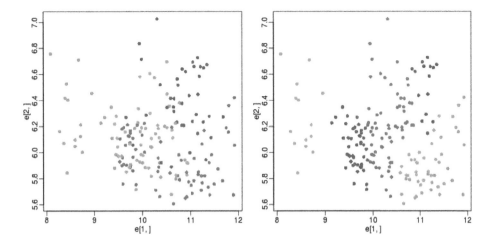

FIGURE 9.4
Plot of gene expression for first two genes (order of appearance in data) with color representing tissue (left) and clusters found with kmeans (right).

```
table(true=tissue,cluster=km$cluster)
```

```
##              cluster
## true            1  2  3  4  5  6  7
##   cerebellum    0  1  8  0  6  0 23
##   colon         2 11  2 15  4  0  0
##   endometrium   0  3  4  0  0  0  8
##   hippocampus  19  0  2  0 10  0  0
##   kidney        7  8 20  0  0  0  4
##   liver         0  0  0  0  0 18  8
##   placenta      0  4  0  0  0  0  2
```

This is very likely due to the fact that the first two genes are not informative regarding tissue type. We can see this in the first plot above. If we instead perform k-means clustering using all of the genes, we obtain a much improved result. To visualize this, we can use an MDS plot:

```
km <- kmeans(t(e), centers=7)
mds <- cmdscale(d)

mypar(1,2)
plot(mds[,1], mds[,2])
plot(mds[,1], mds[,2], col=km$cluster, pch=16)
```

By tabulating the results, we see that we obtain a similar answer to that obtained with hierarchical clustering.

```
table(true=tissue,cluster=km$cluster)
```

```
##              cluster
## true            1  2  3  4  5  6  7
##   cerebellum    0  0  5  0 31  2  0
```

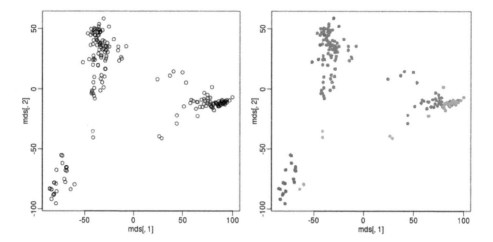

FIGURE 9.5
Plot of gene expression for first two PCs with color representing tissues (left) and clusters found using all genes (right).

```
##    colon        0 34  0  0  0  0  0
##    endometrium  0 15  0  0  0  0  0
##    hippocampus  0  0 31  0  0  0  0
##    kidney       0 37  0  0  0  2  0
##    liver        2  0  0  0  0  0 24
##    placenta     0  0  0  6  0  0  0
```

Heatmaps Heatmaps are ubiquitous in the genomics literature. They are very useful plots for visualizing the measurements for a subset of rows over all the samples. A *dendrogram* is added on top and on the side that is created with hierarchical clustering. We will demonstrate how to create heatmaps from within R. Let's begin by defining a color palette:

```
library(RColorBrewer)
hmcol <- colorRampPalette(brewer.pal(9, "GnBu"))(100)
```

Now, pick the genes with the top variance over all samples:

```
library(genefilter)
rv <- rowVars(e)
idx <- order(-rv)[1:40]
```

While a `heatmap` function is included in R, we recommend the `heatmap.2` function from the `gplots` package on CRAN because it is a bit more customized. For example, it stretches to fill the window. Here we add colors to indicate the tissue on the top:

```
library(gplots) ##Available from CRAN
cols <- palette(brewer.pal(8, "Dark2"))[as.fumeric(tissue)]
head(cbind(colnames(e),cols))
```

```
##                              cols
## [1,] "GSM11805.CEL.gz" "#1B9E77"
## [2,] "GSM11814.CEL.gz" "#1B9E77"
```

```
## [3,] "GSM11823.CEL.gz" "#1B9E77"
## [4,] "GSM11830.CEL.gz" "#1B9E77"
## [5,] "GSM12067.CEL.gz" "#1B9E77"
## [6,] "GSM12075.CEL.gz" "#1B9E77"
```

```
heatmap.2(e[idx,], labCol=tissue,
          trace="none",
          ColSideColors=cols,
          col=hmcol)
```

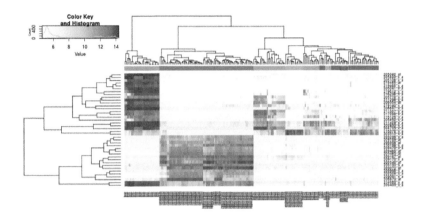

FIGURE 9.6
Heatmap created using the 40 most variable genes and the function heatmap.2.

We did not use tissue information to create this heatmap, and we can quickly see, with just 40 genes, good separation across tissues. {pagebreak}

9.2 Exercises

1. Create a random matrix with no correlation in the following way:

```
set.seed(1)
m = 10000
n = 24
x = matrix(rnorm(m*n),m,n)
colnames(x)=1:n
```

Run hierarchical clustering on this data with the `hclust` function with default parameters to cluster the columns. Create a dendrogram.
In the dendrogram, which pairs of samples are the furthest away from each other?

- A) 7 and 23

- B) 19 and 14

- C) 1 and 16

- D) 17 and 18

2. Set the seed at 1, `set.seed(1)` and replicate the creation of this matrix:

```
m = 10000
n = 24
x = matrix(rnorm(m*n),m,n)
```

then perform hierarchical clustering as in the solution to exercise 1, and find the number of clusters if you use `cuttree` at height 143. This number is a random variable.

Based on the Monte Carlo simulation, what is the standard error of this random variable?

3. Run `kmeans` with 4 centers for the blood RNA data:

```
library(GSE5859Subset)
data(GSE5859Subset)
```

Set the seed to 10, `set.seed(10)` right before running `kmeans` with 5 centers.

Explore the relationship of clusters and information in `sampleInfo`. Which of the following best describes what you find?

- A) `sampleInfo$group` is driving the clusters as the 0s and 1s are in completely different clusters.

- B) The year is driving the clusters.

- C) Date is driving the clusters.

- D) The clusters don't depend on any of the columns of `sampleInfo`

4. Load the data:

```
library(GSE5859Subset)
data(GSE5859Subset)
```

Pick the 25 genes with the highest across sample variance. This function might help:

```
install.packages("matrixStats")
library(matrixStats)
?rowMads ##we use mads due to a outlier sample
```

Use `heatmap.2` to make a heatmap showing the `sampleInfo$group` with color, the date as labels, the rows labelled with chromosome, and scaling the rows.

What do we learn from this heatmap?

- A) The data appears as if it was generated by `rnorm`.

- B) Some genes in chr1 are very variable.

- C) A group of chrY genes are higher in group 0 and appear to drive the clustering. Within those clusters there appears to be clustering by month.

- D) A group of chrY genes are higher in October compared to June and appear to drive the clustering. Within those clusters there appears to be clustering by `samplInfo$group`.

5. Create a large dataset of random data that is completely independent of `sampleInfo$group` like this:

```
set.seed(17)
m = nrow(geneExpression)
n = ncol(geneExpression)
x = matrix(rnorm(m*n),m,n)
g = factor(sampleInfo$g )
```

Create two heatmaps with these data. Show the group g either with labels or colors. First, take the 50 genes with smallest p-values obtained with `rowttests`. Then, take the 50 genes with largest standard deviations.

Which of the following statements is true?

- A) There is no relationship between g and x, but with 8,793 tests some will appear significant by chance. Selecting genes with the t-test gives us a deceiving result.

- B) These two techniques produced similar heatmaps.

- C) Selecting genes with the t-test is a better technique since it permits us to detect the two groups. It appears to find hidden signals.

- D) The genes with the largest standard deviation add variability to the plot and do not let us find the differences between the two groups.

9.3 Conditional Probabilities and Expectations

Prediction problems can be divided into categorical and continuous outcomes. However, many of the algorithms can be applied to both due to the connection between *conditional probabilities* and *conditional expectations*.

For categorical data, for example binary outcomes, if we know the probability of Y being any of the possible outcomes k given a set of predictors $X = (X_1, \ldots, X_p)^\top$,

$$f_k(x) = \Pr(Y = k \mid X = x)$$

we can optimize our predictions. Specifically, for any x we predict the k that has the largest probability $f_k(x)$.

To simplify the exposition below, we will consider the case of binary data. You can think of the probability $\Pr(Y = 1 \mid X = x)$ as the proportion of 1s in the stratum of the population for which $X = x$. Given that the expectation is the average of all Y values, in this case the expectation is equivalent to the probability: $f(x) \equiv \mathrm{E}(Y \mid X = x) = \Pr(Y = 1 \mid X = x)$. We therefore use only the expectation in the descriptions below as it is more general.

In general, the expected value has an attractive mathematical property, which is that it minimizes the expected distance between the predictor $\hat{Y}$ and Y:

$$\mathrm{E}\{(\hat{Y} - Y)^2 \mid X = x\}$$

Regression in the context of prediction We use the son and father height example to illustrate how regression can be interpreted as a machine learning technique. In our example, we are trying to predict the son's height Y based on the father's X. Here we have only one predictor. Now if we were asked to predict the height of a randomly selected son, we would go with the average height:

```
library(rafalib)
mypar(1,1)
data(father.son,package="UsingR")
x=round(father.son$fheight) ##round to nearest inch
y=round(father.son$sheight)
hist(y,breaks=seq(min(y),max(y)))
abline(v=mean(y),col="red",lwd=2)
```

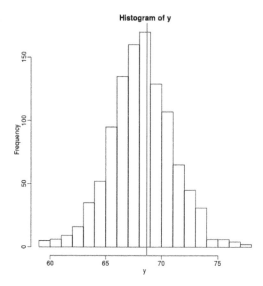

FIGURE 9.7
Histogram of son heights.

In this case, we can also approximate the distribution of Y as normal, which implies the mean maximizes the probability density.

Let's imagine that we are given more information. We are told that the father of this randomly selected son has a height of 71 inches (1.25 SDs taller than the average). What is our prediction now?

```
mypar(1,2)
plot(x,y,xlab="Father's height in inches",ylab="Son's height in inches",
    main=paste("correlation =",signif(cor(x,y),2)))
abline(v=c(-0.35,0.35)+71,col="red")
hist(y[x==71],xlab="Heights",nc=8,main="",xlim=range(y))
```

The best guess is still the expectation, but our strata has changed from all the data, to only the Y with $X = 71$. So we can stratify and take the average, which is the conditional expectation. Our prediction for any x is therefore:

$$f(x) = E(Y \mid X = x)$$

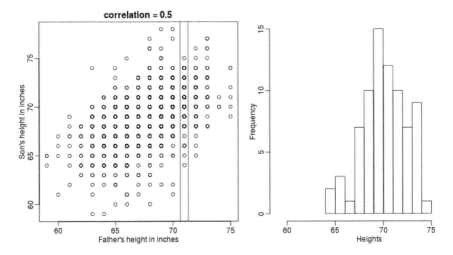

FIGURE 9.8
Son versus father height (left) with the red lines denoting the stratum defined by conditioning on fathers being 71 inches tall. Conditional distribution: son height distribution of stratum defined by 71 inch fathers.

It turns out that because this data is approximated by a bivariate normal distribution, using calculus, we can show that:

$$f(x) = \mu_Y + \rho \frac{\sigma_Y}{\sigma_X}(X - \mu_X)$$

and if we estimate these five parameters from the sample, we get the regression line:

```
mypar(1,2)
plot(x,y,xlab="Father's height in inches",ylab="Son's height in inches",
    main=paste("correlation =",signif(cor(x,y),2)))
abline(v=c(-0.35,0.35)+71,col="red")

fit <- lm(y~x)
abline(fit,col=1)

hist(y[x==71],xlab="Heights",nc=8,main="",xlim=range(y))
abline(v = fit$coef[1] + fit$coef[2]*71, col=1)
```

In this particular case, the regression line provides an optimal prediction function for Y. But this is not generally true because, in the typical machine learning problems, the optimal $f(x)$ is rarely a simple line.

9.4 Exercises

Throughout these exercises it will be useful to remember that when our data are 0s and 1s, probabilities and expectations are the same thing. We can do the math, but here is some R code:

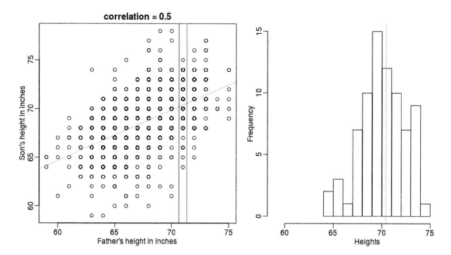

FIGURE 9.9
Son versus father height showing predicted heights based on regression line (left). Conditional distribution with vertical line representing regression prediction.

```
n = 1000
y = rbinom(n,1,0.25)
##proportion of ones Pr(Y)
sum(y==1)/length(y)
##expectaion of Y
mean(y)
```

1. Generate some random data to imitate heights for men (0) and women (1):

```
n = 10000
set.seed(1)
men = rnorm(n,176,7) #height in centimeters
women = rnorm(n,162,7) #height in centimeters
y = c(rep(0,n),rep(1,n))
x = round(c(men,women))
##mix it up
ind = sample(seq(along=y))
y = y[ind]
x = x[ind]
```

1. Using the data generated above, what is the $E(Y|X = 176)$?

2. Now make a plot of $E(Y|X = x)$ for x=seq(160,178) using the data generated in exercise 1.

If you are predicting female or male based on height and want your probability of success to be larger than 0.5, what is the largest height where you predict female?

9.5   Smoothing

Smoothing is a very powerful technique used all across data analysis. It is designed to estimate $f(x)$ when the shape is unknown, but assumed to be *smooth*. The general idea is to group data points that are expected to have similar expectations and compute the average, or fit a simple parametric model. We illustrate two smoothing techniques using a gene expression example.

The following data are gene expression measurements from replicated RNA samples.

```
##Following three packages are available from Bioconductor
library(Biobase)
library(SpikeIn)
library(hgu95acdf)

## Warning: replacing previous import by 'utils::tail' when loading
## 'hgu95acdf'

## Warning: replacing previous import by 'utils::head' when loading
## 'hgu95acdf'

data(SpikeIn95)
```

We consider the data used in an MA-plot comparing two replicated samples ($Y = $ log ratios and $X = $ averages) and take down-sample in a way that balances the number of points for different strata of X (code not shown):

```
library(rafalib)
mypar()
plot(X,Y)
```

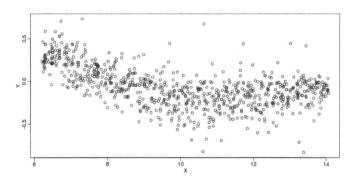

FIGURE 9.10
MA-plot comparing gene expression from two arrays.

In the MA plot we see that Y depends on X. This dependence must be a bias because these are based on replicates, which means Y should be 0 on average regardless of X. We want to predict $f(x) = \mathrm{E}(Y \mid X = x)$ so that we can remove this bias. Linear regression does not capture the apparent curvature in $f(x)$:

```
mypar()
plot(X,Y)
fit <- lm(Y~X)
points(X,Y,pch=21,bg=ifelse(Y>fit$fitted,1,3))
abline(fit,col=2,lwd=4,lty=2)
```

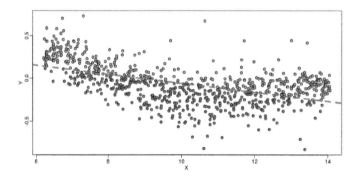

FIGURE 9.11

MA-plot comparing gene expression from two arrays with fitted regression line. The two colors represent positive and negative residuals.

The points above the fitted line (green) and those below (purple) are not evenly distributed. We therefore need an alternative more flexible approach.

9.6   Bin Smoothing

Instead of fitting a line, let's go back to the idea of stratifying and computing the mean. This is referred to as *bin smoothing*. The general idea is that the underlying curve is "smooth" enough so that, in small bins, the curve is approximately constant. If we assume the curve is constant, then all the Y in that bin have the same expected value. For example, in the plot below, we highlight points in a bin centered at 8.6, as well as the points of a bin centered at 12.1, if we use bins of size 1. We also show the fitted mean values for the Y in those bins with dashed lines (code not shown):

By computing this mean for bins around every point, we form an estimate of the underlying curve $f(x)$. Below we show the procedure happening as we move from the smallest value of x to the largest. We show 10 intermediate cases as well (code not shown):

The final result looks like this (code not shown):

There are several functions in R that implement bin smoothers. One example is `ksmooth`. However, in practice, we typically prefer methods that use slightly more complicated models than fitting a constant. The final result above, for example, is somewhat wiggly. Methods such as `loess`, which we explain next, improve on this.

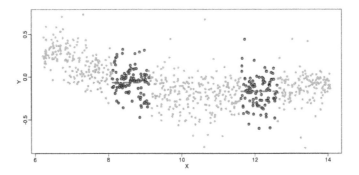

FIGURE 9.12
MAplot comparing gene expression from two arrays with bin smoother fit shown for two points.

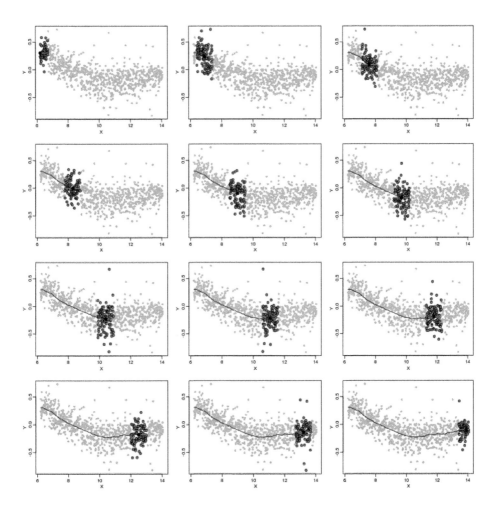

FIGURE 9.13
Illustration of how bin smoothing estimates a curve. Showing 12 steps of process.

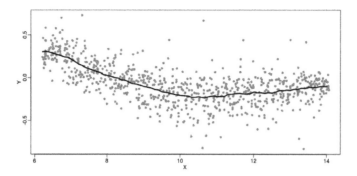

FIGURE 9.14
MA-plot with curve obtained with bin-smoothed curve shown.

9.7 Loess

Local weighted regression (loess) is similar to bin smoothing in principle. The main difference is that we approximate the local behavior with a line or a parabola. This permits us to expand the bin sizes, which stabilizes the estimates. Below we see lines fitted to two bins that are slightly larger than those we used for the bin smoother (code not shown). We can use larger bins because fitting lines provide slightly more flexibility.

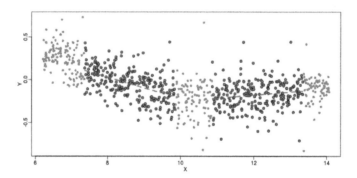

FIGURE 9.15
MA-plot comparing gene expression from two arrays with bin local regression fit shown for two points.

As we did for the bin smoother, we show 12 steps of the process that leads to a loess fit (code not shown):

The final result is a smoother fit than the bin smoother since we use larger sample sizes to estimate our local parameters (code not shown):

The function `loess` performs this analysis for us:

```
fit <- loess(Y~X, degree=1, span=1/3)

newx <- seq(min(X),max(X),len=100)
smooth <- predict(fit,newdata=data.frame(X=newx))
```

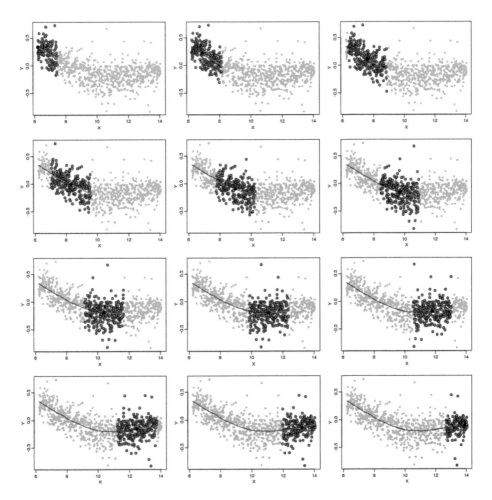

FIGURE 9.16
Illustration of how loess estimates a curve. Showing 12 steps of the process.

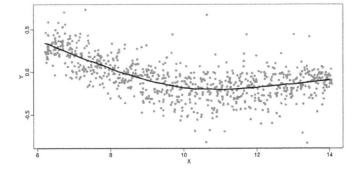

FIGURE 9.17
MA-plot with curve obtained with loess.

```
mypar ()
plot(X,Y,col="darkgrey",pch=16)
lines(newx,smooth,col="black",lwd=3)
```

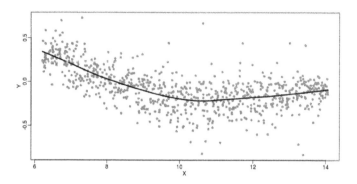

FIGURE 9.18
Loess fitted with the loess function.

There are three other important differences between `loess` and the typical bin smoother. The first is that rather than keeping the bin size the same, `loess` keeps the number of points used in the local fit the same. This number is controlled via the `span` argument which expects a proportion. For example, if `N` is the number of data points and `span=0.5`, then for a given x , `loess` will use the `0.5*N` closest points to x for the fit. The second difference is that, when fitting the parametric model to obtain $f(x)$, `loess` uses weighted least squares, with higher weights for points that are closer to x. The third difference is that `loess` has the option of fitting the local model robustly. An iterative algorithm is implemented in which, after fitting a model in one iteration, outliers are detected and downweighted for the next iteration. To use this option, we use the argument `family="symmetric"`.

9.8   Exercises

1.   Generate the following data:

```
n = 10000
set.seed(1)
men = rnorm(n,176,7) #height in centimeters
women = rnorm(n,162,7) #height in centimeters
y = c(rep(0,n),rep(1,n))
x = round(c(men,women))
##mix it up
ind = sample(seq(along=y))
y = y[ind]
x = x[ind]
```

Set the seed at 5, `set.seed(5)` and take a random sample of 250 from:

```
set.seed(5)
```

```
N = 250
ind = sample(length(y),N)
Y = y[ind]
X = x[ind]
```

Use loess to estimate $f(x) = E(Y|X = x)$ using the default parameters. What is the predicted $f(168)$?

2. The loess estimate above is a random variable. We can compute standard errors for it. Here we use Monte Carlo to demonstrate that it is a random variable. Use Monte Carlo simulation to estimate the standard error of your estimate of $f(168)$.

Set the seed to 5, `set.seed(5)` and perform 10,000 simulations and report the SE of the loess based estimate.

9.9   Class Prediction

Here we give a brief introduction to the main task of machine learning: class prediction. In fact, many refer to class prediction as machine learning and we sometimes use the two terms interchangeably. We give a very brief introduction to this vast topic, focusing on some specific examples.

Some of the examples we give here are motivated by those in the excellent textbook *The Elements of Statistical Learning: Data Mining, Inference, and Prediction*, by Trevor Hastie, Robert Tibshirani and Jerome Friedman, which can be found here[2].

Similar to inference in the context of regression, Machine Learning (ML) studies the relationships between outcomes Y and covariates X. In ML, we call X the predictors or features. The main difference between ML and inference is that, in ML, we are interested mainly in predicting Y using X. Statistical models are used, but while in inference we estimate and interpret model parameters, in ML they are mainly a means to an end: predicting Y.

Here we introduce the main concepts needed to understand ML, along with two specific algorithms: regression and k nearest neighbors (kNN). Keep in mind that there are dozens of popular algorithms that we do not cover here.

In a previous section, we covered the very simple one-predictor case. However, most of ML is concerned with cases with more than one predictor. For illustration purposes, we move to a case in which X is two dimensional and Y is binary. We simulate a situation with a non-linear relationship using an example from the Hastie, Tibshirani and Friedman book. In the plot below, we show the actual values of $f(x_1, x_2) = E(Y \mid X_1 = x_1, X_2 = x_2)$ using colors. The following code is used to create a relatively complex conditional probability function. We create the test and train data we use later (code not shown). Here is the plot of $f(x_1, x_2)$ with red representing values close to 1, blue representing values close to 0, and yellow values in between.

If we show points for which $E(Y \mid X = x) > 0.5$ in red and the rest in blue, we see the boundary region that denotes the boundary in which we switch from predicting 0 to 1.

The above plots relate to the "truth" that we do not get to see. Most ML methodology is concerned with estimating $f(x)$. A typical first step is usually to consider a sample, referred to as the training set, to estimate $f(x)$. We will review two specific ML techniques. First, we need to review the main concept we use to evaluate the performance of these methods.

---

[2]http://statweb.stanford.edu/~tibs/ElemStatLearn/

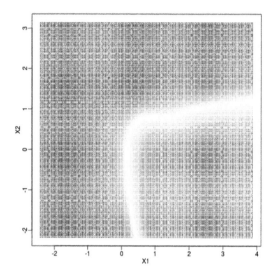

FIGURE 9.19
Probability of Y=1 as a function of X1 and X2. Red is close to 1, yellow close to 0.5, and blue close to 0.

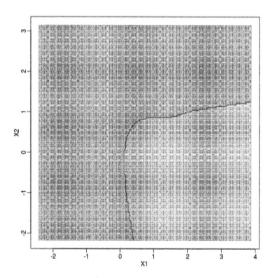

FIGURE 9.20
Bayes rule. The line divides part of the space for which probability is larger than 0.5 (red) and lower than 0.5 (blue).

Training and test sets In the code (not shown) for the first plot in this chapter, we created a test and a training set. We plot them here:

```
#x, test, cols, and coltest were created in code that was not shown
#x is training x1 and x2, test is test x1 and x2
#cols (0=blue, 1=red) are training observations
#coltests are test observations
mypar(1,2)
plot(x,pch=21,bg=cols,xlab="X1",ylab="X2",xlim=XLIM,ylim=YLIM)
plot(test,pch=21,bg=colstest,xlab="X1",ylab="X2",xlim=XLIM,ylim=YLIM)
```

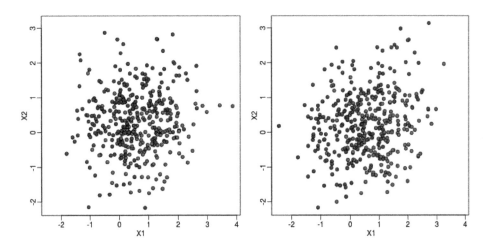

FIGURE 9.21
Training data (left) and test data (right).

You will notice that the test and train set have similar global properties since they were generated by the same random variables (more blue towards the bottom right), but are, by construction, different. The reason we create test and training sets is to detect over-training by testing on a different data than the one used to fit models or train algorithms. We will see how important this is below.

Predicting with regression A first naive approach to this ML problem is to fit a two variable linear regression model:

```
##x and y were created in the code (not shown) for the first plot
#y is outcome for the training set
X1 <- x[,1] ##these are the covariates
X2 <- x[,2]
fit1 <- lm(y~X1+X2)
```

Once we the have fitted values, we can estimate $f(x_1, x_2)$ with $\hat{f}(x_1, x_2) = \hat{\beta}_0 + \hat{\beta}_1 x_1 + \hat{\beta}_2 x_2$. To provide an actual prediction, we simply predict 1 when $\hat{f}(x_1, x_2) > 0.5$. We now examine the error rates in the test and training sets and also plot the boundary region:

```
##prediction on train
yhat <- predict(fit1)
```

```
yhat <- as.numeric(yhat>0.5)
cat("Linear regression prediction error in train:",1-mean(yhat==y),"\n")
```

Linear regression prediction error in train: 0.295

We can quickly obtain predicted values for any set of values using the **predict** function:

```
yhat <- predict(fit1,newdata=data.frame(X1=newx[,1],X2=newx[,2]))
```

Now we can create a plot showing where we predict 1s and where we predict 0s, as well as the boundary. We can also use the **predict** function to obtain predicted values for our test set. Note that nowhere do we fit the model on the test set:

```
colshat <- yhat
colshat[yhat>=0.5] <- mycols[2]
colshat[yhat<0.5] <- mycols[1]
m <- -fit1$coef[2]/fit1$coef[3] #boundary slope
b <- (0.5 - fit1$coef[1])/fit1$coef[3] #boundary intercept

##prediction on test
yhat <- predict(fit1,newdata=data.frame(X1=test[,1],X2=test[,2]))
yhat <- as.numeric(yhat>0.5)
cat("Linear regression prediction error in test:",1-mean(yhat==ytest),"\n")
```

Linear regression prediction error in test: 0.3075

```
plot(test,type="n",xlab="X1",ylab="X2",xlim=XLIM,ylim=YLIM)
abline(b,m)
points(newx,col=colshat,pch=16,cex=0.35)

##test was created in the code (not shown) for the first plot
points(test,bg=cols,pch=21)
```

The error rates in the test and train sets are quite similar. Thus, we do not seem to be over-training. This is not surprising as we are fitting a 2 parameter model to 400 data points. However, note that the boundary is a line. Because we are fitting a plane to the data, there is no other option here. The linear regression method is too rigid. The rigidity makes it stable and avoids over-training, but it also keeps the model from adapting to the non-linear relationship between Y and X. We saw this before in the smoothing section. The next ML technique we consider is similar to the smoothing techniques described before.

K-nearest neighbor K-nearest neighbors (kNN) is similar to bin smoothing, but it is easier to adapt to multiple dimensions. Basically, for any point x for which we want an estimate, we look for the k nearest points and then take an average of these points. This gives us an estimate of $f(x_1, x_2)$, just like the bin smoother gave us an estimate of a curve. We can now control flexibility through k. Here we compare $k = 1$ and $k = 100$.

```
library(class)
mypar(2,2)
for(k in c(1,100)){
  ##predict on train
  yhat <- knn(x,x,y,k=k)
  cat("KNN prediction error in train:",1-mean((as.numeric(yhat)-1)==y),"\n")
  ##make plot
```

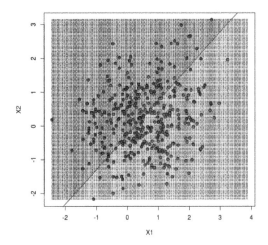

FIGURE 9.22
We estimate the probability of 1 with a linear regression model with X1 and X2 as predictors. The resulting prediction map is divided into parts that are larger than 0.5 (red) and lower than 0.5 (blue).

```
  yhat <- knn(x,test,y,k=k)
    cat("KNN prediction error in test:",1-mean((as.numeric(yhat)-1)==ytest),"\n")
}
```

```
## KNN prediction error in train: 0
## KNN prediction error in test: 0.375
## KNN prediction error in train: 0.2425
## KNN prediction error in test: 0.2825
```

To visualize why we make no errors in the train set and many errors in the test set when $k = 1$ and obtain more stable results from $k = 100$, we show the prediction regions (code not shown):

When $k = 1$, we make no mistakes in the training test since every point is its closest neighbor and it is equal to itself. However, we see some islands of blue in the red area that, once we move to the test set, are more error prone. In the case $k = 100$, we do not have this problem and we also see that we improve the error rate over linear regression. We can also see that our estimate of $f(x_1, x_2)$ is closer to the truth.

Bayes rule Here we include a comparison of the test and train set errors for various values of k. We also include the error rate that we would make if we actually knew $E(Y \mid X_1 = x1, X_2 = x_2)$ referred to as *Bayes Rule*.

We start by computing the error rates...

```
###Bayes Rule
yhat <- apply(test,1,p)
cat("Bayes rule prediction error in train",1-mean(round(yhat)==y),"\n")
```

```
## Bayes rule prediction error in train 0.2775
```

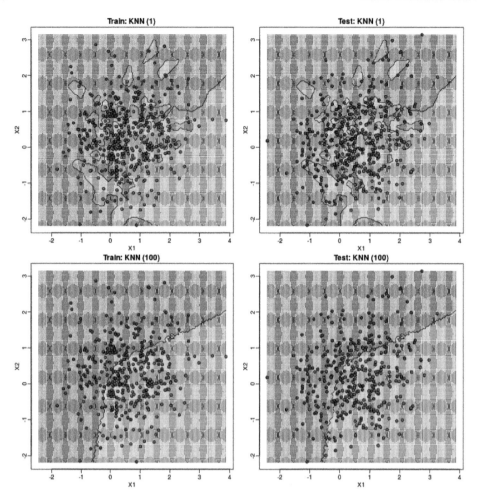

FIGURE 9.23

Prediction regions obtained with kNN for k=1 (top) and k=200 (bottom). We show both train (left) and test data (right).

```
bayes.error=1-mean(round(yhat)==y)
train.error <- rep(0,16)
test.error <- rep(0,16)
for(k in seq(along=train.error)){
  ##predict on train
  yhat <- knn(x,x,y,k=2^(k/2))
  train.error[k] <- 1-mean((as.numeric(yhat)-1)==y)
  ##prediction on test
  yhat <- knn(x,test,y,k=2^(k/2))
  test.error[k] <- 1-mean((as.numeric(yhat)-1)==y)
}
```

... and then plot the error rates against values of k. We also show the Bayes rules error rate as a horizontal line.

```
ks <- 2^(seq(along=train.error)/2)
mypar()
```

```
plot(ks,train.error,type="n",xlab="K",ylab="Prediction Error",log="x",
     ylim=range(c(test.error,train.error)))
lines(ks,train.error,type="b",col=4,lty=2,lwd=2)
lines(ks,test.error,type="b",col=5,lty=3,lwd=2)
abline(h=bayes.error,col=6)
legend("bottomright",c("Train","Test","Bayes"),col=c(4,5,6),lty=c(2,3,1),box.lwd=0)
```

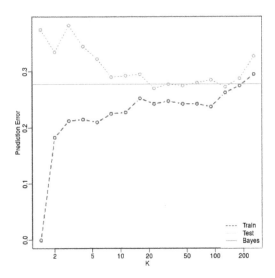

FIGURE 9.24
Prediction error in train (pink) and test (green) versus number of neighbors. The yellow line represents what one obtains with Bayes Rule.

Note that these error rates are random variables and have standard errors. In the next section we describe cross-validation which helps reduce some of this variability. However, even with this variability, the plot clearly shows the problem of over-fitting when using values lower than 20 and under-fitting with values above 100.

9.10 Cross-validation

Here we describe *cross-validation*: one of the fundamental methods in machine learning for method assessment and picking parameters in a prediction or machine learning task. Suppose we have a set of observations with many features and each observation is associated with a label. We will call this set our training data. Our task is to predict the label of any new samples by learning patterns from the training data. For a concrete example, let's consider gene expression values, where each gene acts as a feature. We will be given a new set of unlabeled data (the test data) with the task of predicting the tissue type of the new samples.

If we choose a machine learning algorithm with a tunable parameter, we have to come up with a strategy for picking an optimal value for this parameter. We could try some values, and then just choose the one which performs the best on our training data, in terms of the number of errors the algorithm would make if we apply it to the samples we have been given for training. However, we have seen how this leads to over-fitting.

Let's start by loading the tissue gene expression dataset:

```
library(tissuesGeneExpression)
data(tissuesGeneExpression)
```

For illustration purposes, we will drop one of the tissues which doesn't have many samples:

```
table(tissue)
```

```
## tissue
##   cerebellum       colon endometrium hippocampus      kidney       liver
##           38          34          15          31          39          26
##     placenta
##            6
```

```
ind <- which(tissue != "placenta")
y <- tissue[ind]
X <- t( e[,ind] )
```

This tissue will not form part of our example.

Now let's try out k-nearest neighbors for classification, using $k = 5$. What is our average error in predicting the tissue in the training set, when we've used the same data for training and for testing?

```
library(class)
pred <- knn(train =  X, test = X, cl=y, k=5)
mean(y != pred)
```

```
## [1] 0
```

We have no errors in prediction in the training set with $k = 5$. What if we use $k = 1$?

```
pred <- knn(train=X, test=X, cl=y, k=1)
mean(y != pred)
```

```
## [1] 0
```

Trying to classify the same observations as we use to *train* the model can be very misleading. In fact, for k-nearest neighbors, using k=1 will always give 0 classification error in the training set, because we use the single observation to classify itself. The reliable way to get a sense of the performance of an algorithm is to make it give a prediction for a sample it has never seen. Similarly, if we want to know what the best value for a tunable parameter is, we need to see how different values of the parameter perform on samples, which are not in the training data.

Cross-validation is a widely-used method in machine learning, which solves this training and test data problem, while still using all the data for testing the predictive accuracy. It accomplishes this by splitting the data into a number of *folds*. If we have N folds, then the first step of the algorithm is to train the algorithm using $(N-1)$ of the folds, and test the algorithm's accuracy on the single left-out fold. This is then repeated N times until each fold has been used as in the *test* set. If we have M parameter settings to try out, then this is accomplished in an outer loop, so we have to fit the algorithm a total of $N \times M$ times.

We will use the **createFolds** function from the **caret** package to make 5 folds of our gene expression data, which are balanced over the tissues. Don't be confused by the fact that the `createFolds` function uses the same letter 'k' as the 'k' in k-nearest neighbors. These 'k' are totally unrelated. The caret function `createFolds` is asking for how many folds to create, the N from above. The 'k' in the `knn` function is for how many closest observations to use in classifying a new sample. Here we will create 10 folds:

```
library(caret)
set.seed(1)
idx <- createFolds(y, k=10)
sapply(idx, length)

## Fold01 Fold02 Fold03 Fold04 Fold05 Fold06 Fold07 Fold08 Fold09 Fold10
##     18     19     17     17     18     20     19     19     20     16
```

The folds are returned as a list of numeric indices. The first fold of data is therefore:

```
y[idx[[1]]] ##the labels

##  [1] "kidney"      "kidney"      "hippocampus" "hippocampus" "hippocampus"
##  [6] "cerebellum"  "cerebellum"  "cerebellum"  "colon"       "colon"
## [11] "colon"       "colon"       "kidney"      "kidney"      "endometrium"
## [16] "endometrium" "liver"       "liver"
```

```
head( X[idx[[1]], 1:3] ) ##the genes (only showing the first 3 genes...)

##                   1007_s_at  1053_at  117_at
## GSM12075.CEL.gz    9.966782 6.060069 7.644452
## GSM12098.CEL.gz    9.945652 5.927861 7.847192
## GSM21214.cel.gz   10.955428 5.776781 7.493743
## GSM21218.cel.gz   10.757734 5.984170 8.525524
## GSM21230.cel.gz   11.496114 5.760156 7.787561
## GSM87086.cel.gz    9.798633 5.862426 7.279199
```

We can see that, in fact, the tissues are fairly equally represented across the 10 folds:

```
sapply(idx, function(i) table(y[i]))
```

##	Fold01	Fold02	Fold03	Fold04	Fold05	Fold06	Fold07	Fold08	Fold09
## cerebellum	3	4	4	4	4	4	4	4	4
## colon	4	3	3	3	4	4	3	3	4
## endometrium	2	2	1	1	1	2	1	2	2
## hippocampus	3	3	3	3	3	3	4	3	3
## kidney	4	4	3	4	4	4	4	4	4
## liver	2	3	3	2	2	3	3	3	3

##	Fold10
## cerebellum	3
## colon	3
## endometrium	1
## hippocampus	3
## kidney	4
## liver	2

Because tissues have very different gene expression profiles, predicting tissue with all genes will be very easy. For illustration purposes we will try to predict tissue type with just two dimensional data. We will reduce the dimension of our data using `cmdscale`:

```
library(rafalib)
mypar()
Xsmall <- cmdscale(dist(X))
plot(Xsmall,col=as.fumeric(y))
legend("topleft",levels(factor(y)),fill=seq_along(levels(factor(y))))
```

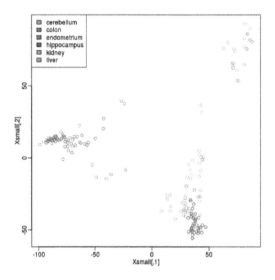

FIGURE 9.25
First two PCs of the tissue gene expression data with color representing tissue. We use these two PCs as our two predictors throughout.

Now we can try out the k-nearest neighbors method on a single fold. We provide the **knn** function with all the samples in **Xsmall** *except* those which are in the first fold. We remove these samples using the code **-idx[[1]]** inside the square brackets. We then use those samples in the test set. The **cl** argument is for the true classifications or labels (here, tissue) of the training data. We use 5 observations to classify in our k-nearest neighbor algorithm:

```
pred <- knn(train=Xsmall[ -idx[[1]] , ], test=Xsmall[ idx[[1]], ], cl=y[ -idx[[1]] ], k=5)
table(true=y[ idx[[1]] ], pred)
```

```
##                pred
## true          cerebellum colon endometrium hippocampus kidney liver
##    cerebellum          2     0           0           1      0     0
##    colon               0     4           0           0      0     0
##    endometrium         0     0           1           0      1     0
##    hippocampus         1     0           0           2      0     0
##    kidney              0     0           0           0      4     0
##    liver               0     0           0           0      0     2
```

```
mean(y[ idx[[1]] ] != pred)
```

```
## [1] 0.1666667
```

Now we have some misclassifications. How well do we do for the rest of the folds?

```
for (i in 1:10) {
  pred <- knn(train=Xsmall[ -idx[[i]] , ], test=Xsmall[ idx[[i]], ], cl=y[ -idx[[i]] ], k=5)
  print(paste0(i,") error rate: ", round(mean(y[ idx[[i]] ] != pred),3)))
}
```

```
## [1] "1) error rate: 0.167"
## [1] "2) error rate: 0.105"
## [1] "3) error rate: 0.118"
## [1] "4) error rate: 0.118"
## [1] "5) error rate: 0.278"
## [1] "6) error rate: 0.05"
## [1] "7) error rate: 0.105"
## [1] "8) error rate: 0.211"
## [1] "9) error rate: 0.15"
## [1] "10) error rate: 0.312"
```

So we can see there is some variation for each fold, with error rates hovering around 0.1-0.3. But is k=5 the best setting for the k parameter? In order to explore the best setting for k, we need to create an outer loop, where we try different values for k, and then calculate the average test set error across all the folds.

We will try out each value of k from 1 to 12. Instead of using two `for` loops, we will use `sapply`:

```
set.seed(1)
ks <- 1:12
res <- sapply(ks, function(k) {
  ##try out each version of k from 1 to 12
  res.k <- sapply(seq_along(idx), function(i) {
    ##loop over each of the 10 cross-validation folds
    ##predict the held-out samples using k nearest neighbors
    pred <- knn(train=Xsmall[ -idx[[i]], ],
                test=Xsmall[ idx[[i]], ],
                cl=y[ -idx[[i]] ], k = k)
    ##the ratio of misclassified samples
    mean(y[ idx[[i]] ] != pred)
  })
  ##average over the 10 folds
  mean(res.k)
})
```

Now for each value of k, we have an associated test set error rate from the cross-validation procedure.

```
res
```

```
##  [1] 0.1978212 0.1703423 0.1882933 0.1750989 0.1613291 0.1500791 0.1552670
##  [8] 0.1884813 0.1822020 0.1763197 0.1761318 0.1813197
```

We can then plot the error rate for each value of k, which helps us to see in what region there might be a minimal error rate:

```
plot(ks, res, type="o",ylab="misclassification error")
```

Remember, because the training set is a random sample and because our fold-generation procedure involves random number generation, the "best" value of k we pick through this procedure is also a random variable. If we had new training data and if we recreated our folds, we might get a different value for the optimal k.

Finally, to show that gene expression can perfectly predict tissue, we use 5 dimensions instead of 2, which results in perfect prediction:

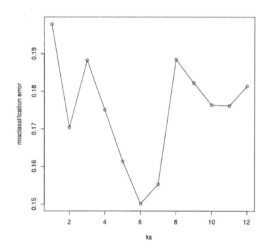

FIGURE 9.26
Misclassification error versus number of neighbors.

```
Xsmall <- cmdscale(dist(X),k=5)
set.seed(1)
ks <- 1:12
res <- sapply(ks, function(k) {
  res.k <- sapply(seq_along(idx), function(i) {
    pred <- knn(train=Xsmall[ -idx[[i]], ],
                test=Xsmall[ idx[[i]], ],
                cl=y[ -idx[[i]] ], k = k)
    mean(y[ idx[[i]] ] != pred)
  })
  mean(res.k)
})
plot(ks, res, type="o",ylim=c(0,0.20),ylab="misclassification error")
```

Important note: we applied `cmdscale` to the entire dataset to create a smaller one for illustration purposes. However, in a real machine learning application, this may result in an underestimation of test set error for small sample sizes, where dimension reduction using the unlabeled full dataset gives a boost in performance. A safer choice would have been to transform the data separately for each fold, by calculating a rotation and dimension reduction using the training set only and applying this to the test set.

9.11   Exercises

Load the following dataset:

```
library(GSE5859Subset)
data(GSE5859Subset)
```

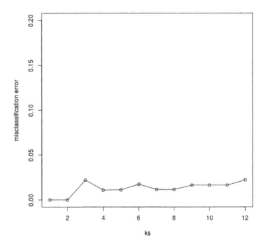

FIGURE 9.27
Misclassification error versus number of neighbors when we use 5 dimensions instead of 2.

And define the outcome and predictors. To make the problem more difficult, we will only consider autosomal genes:

```
y = factor(sampleInfo$group)
X = t(geneExpression)
out = which(geneAnnotation$CHR%in%c("chrX","chrY"))
X = X[,-out]
```

1. Use the `createFold` function in the `caret` package, set the seed to 1 `set.seed(1)` and create 10 folds of y.

   Question: What is the 2nd entry in the fold 3?

2. We are going to use kNN. We are going to consider a smaller set of predictors by using *filtering* genes using t-tests. Specifically, we will perform a t-test and select the m genes with the smallest p-values.

   Let $m = 8$ and $k = 5$ and train kNN by leaving out the second fold `idx[[2]]`. How many mistakes do we make on the test set? Remember it is indispensable that you perform the t-test on the training data.

3. Now run through all 5 folds. What is our error rate?

4. Now we are going to select the best values of k and m. Use the expand grid function to try out the following values:

```
ms=2^c(1:11)
ks=seq(1,9,2)
params = expand.grid(k=ks,m=ms)
```

Now use apply or a for-loop to obtain error rates for each of these pairs of parameters. Which pair of parameters minimizes the error rate?

5. Repeat exercise 4, but now perform the t-test filtering before the cross validation. Note how this biases the entire result and gives us much lower estimated error rates.

6. Repeat exercise 3, but now, instead of `sampleInfo$group` , use

```
y = factor(as.numeric(format( sampleInfo$date, "%m")=="06"))
```

What is the minimum error rate now?

We achieve much lower error rates when predicting date than when predicting the group. Because group is confounded with date, it is very possible that these predictors have no information about group and that our lower 0.5 error rates are due to the confounding with date. We will learn more about this in the batch effect chapter.

10

Batch Effects

One often overlooked complication with high-throughput studies is batch effects, which occur because measurements are affected by laboratory conditions, reagent lots, and personnel differences. This becomes a major problem when batch effects are confounded with an outcome of interest and lead to incorrect conclusions. In this chapter, we describe batch effects in detail: how to detect, interpret, model, and adjust for batch effects.

Batch effects are the biggest challenge faced by genomics research, especially in the context of precision medicine. The presence of batch effects in one form or another has been reported among most, if not all, high-throughput technologies [Leek et al. (2010) Nature Reviews Genetics 11, 733-739]. But batch effects are not specific to genomics technology. In fact, in a 1972 paper, WJ Youden describes batch effects in the context of empirical estimates of physical constants. He pointed out the "subjective character of present estimates" of physical constants and how estimates changed from laboratory to laboratory. For example, in Table 1, Youden shows the following estimates of the astronomical unit from different laboratories. The reports included an estimate of spread (what we now would call a confidence interval).

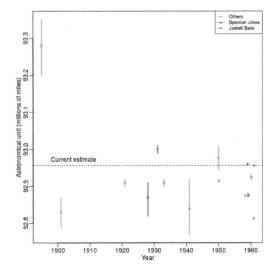

FIGURE 10.1
Estimates of the astronomical unit with estimates of spread, versus year it was reported. The two laboratories that reported more than one estimate are shown in color.

Judging by the variability across labs and the fact that the reported bounds do not explain this variability, clearly shows the presence of an effect that differs across labs, but not within. This type of variability is what we call a batch effect. Note that there are laboratories that reported two estimates (purple and orange) and batch effects are seen across the two different measurements from the same laboratories as well.

We can use statistical notation to precisely describe the problem. The scientists making these measurements assumed they observed:

$$Y_{i,j} = \mu + \varepsilon_{i,j}, j = 1, \ldots, N$$

with $Y_{i,j}$ the j-th measurement of laboratory i, μ the true physical constant, and $\varepsilon_{i,j}$ independent measurement error. To account for the variability introduced by $\varepsilon_{i,j}$, we compute standard errors from the data. As we saw earlier in the book, we estimate the physical constant with the average of the N measurements...

$$\bar{Y}_i = \frac{1}{N} \sum_{i=1}^{N} Y_{i,j}$$

.. and we can construct a confidence interval by:

$$\bar{Y}_i \pm 2s_i/\sqrt{N} \text{ with } s_i^2 = \frac{1}{N-1} \sum_{i=1}^{N} (Y_{i,j} - \bar{Y}_i)^2$$

However, this confidence interval will be too small because it does not catch the batch effect variability. A more appropriate model is:

$$Y_{i,j} = \mu + \gamma_i + \varepsilon_{i,j}, j = 1, \ldots, N$$

with γ_i a laboratory specific bias or *batch effect.*

From the plot it is quite clear that the variability of γ across laboratories is larger than the variability of ε within a lab. The problem here is that there is no information about γ in the data from a single lab. The statistical term for the problem is that μ and γ are unidentifiable. We can estimate the sum $\mu_i + \gamma_i$, but we can't distinguish one from the other.

We can also view γ as a random variable. In this case, each laboratory has an error term γ_i that is the same across measurements from that lab, but different from lab to lab. Under this interpretation the problem is that:

$$s_i/\sqrt{N} \text{ with } s_i^2 = \frac{1}{N-1} \sum_{i=1}^{N} (Y_{ij} - \bar{Y}_i)^2$$

is an underestimate of the standard error since it does not account for the within lab correlation induced by γ.

With data from several laboratories we can in fact estimate the γ, if we assume they average out to 0. Or we can consider them to be random effects and simply estimate a new estimate and standard error with all measurements. Here is a confidence interval treating each reported average as a random observation:

```
avg <- mean(dat[,3])
se <- sd(dat[,3]) / sqrt(nrow(dat))

## 95% confidence interval is: [ 92.8727 , 92.98542 ]

## which does include the current estimate is: 92.95604
```

Youden's paper also includes batch effect examples from more recent estimates of the speed of light, as well as estimates of the gravity constant. Here we demonstrate the widespread presence and complex nature of batch effects in high-throughput biological measurements.

10.1 Confounding

Batch effects have the most devastating effects when they are *confounded* with outcomes of interest. Here we describe confounding and how it relates to data interpretation.

"Correlation is not causation" is one of the most important lessons you should take from this or any other data analysis course. A common example for why this statement is so often true is confounding. Simply stated confounding occurs when we observe a correlation or association between X and Y, but this is strictly the result of both X and Y depending on an extraneous variable Z. Here we describe Simpson's paradox, an example based on a famous legal case, and an example of confounding in high-throughput biology.

Example of Simpson's Paradox Admission data from U.C. Berkeley 1973 showed that more men were being admitted than women: 44% men were admitted compared to 30% women. This actually led to a lawsuit[1]. See: PJ Bickel, EA Hammel, and JW O'Connell. Science (1975). Here is the data:

```
library(dagdata)
data(admissions)
admissions$total=admissions$Percent*admissions$Number/100

##percent men get in
sum(admissions$total[admissions$Gender==1]/sum(admissions$Number[admissions$Gender==1]))
```

```
## [1] 0.4451951
```

```
##percent women get in
sum(admissions$total[admissions$Gender==0]/sum(admissions$Number[admissions$Gender==0]))
```

```
## [1] 0.3033351
```

A chi-square test clearly rejects the hypothesis that gender and admission are independent:

```
##make a 2 x 2 table
index = admissions$Gender==1
men = admissions[index,]
women = admissions[!index,]
menYes = sum(men$Number*men$Percent/100)
menNo = sum(men$Number*(1-men$Percent/100))
womenYes = sum(women$Number*women$Percent/100)
womenNo = sum(women$Number*(1-women$Percent/100))
tab = matrix(c(menYes,womenYes,menNo,womenNo),2,2)
print(chisq.test(tab)$p.val)
```

```
## [1] 9.139492e-22
```

But closer inspection shows a paradoxical result. Here are the percent admissions by major:

```
y=cbind(admissions[1:6,c(1,3)],admissions[7:12,3])
colnames(y)[2:3]=c("Male","Female")
y
```

---

[1]http://en.wikipedia.org/wiki/Simpson%27s_paradox#Berkeley_gender_bias_case

```
##   Major Male Female
## 1     A   62     82
## 2     B   63     68
## 3     C   37     34
## 4     D   33     35
## 5     E   28     24
## 6     F    6      7
```

Notice that we no longer see a clear gender bias. The chi-square test we performed above suggests a dependence between admission and gender. Yet when the data is grouped by major, this dependence seems to disappear. What's going on?

This is an example of *Simpson's paradox*. A plot showing the percentages that applied to a major against the percent that get into that major, for males and females starts to point to an explanation.

```
y=cbind(admissions[1:6,5],admissions[7:12,5])
y=sweep(y,2,colSums(y),"/")*100
x=rowMeans(cbind(admissions[1:6,3],admissions[7:12,3]))

library(rafalib)
mypar()
matplot(x,y,xlab="percent that gets in the major",
        ylab="percent that applies to major",
        col=c("blue","red"),cex=1.5)
legend("topleft",c("Male","Female"),col=c("blue","red"),pch=c("1","2"),box.lty=0)
```

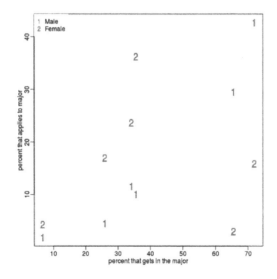

FIGURE 10.2
Percent of students that applied versus percent that were admitted by gender.

What the plot suggests is that males were much more likely to apply to "easy" majors. The plot shows that males and "easy" majors are confounded.

Confounding explained graphically Here we visualize the confounding. In the plots below, each letter represents a person. Accepted individuals are denoted in green and not admitted in orange. The letter indicates the major. In this first plot we group all the students together and notice that the proportion of green is larger for men.

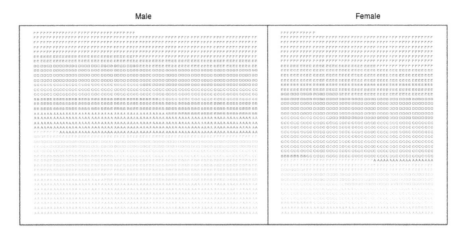

FIGURE 10.3
Admitted are in green and majors are denoted with letters. Here we clearly see that more males were admitted.

Now we stratify the data by major. The key point here is that most of the accepted men (green) come from the easy majors: A and B.

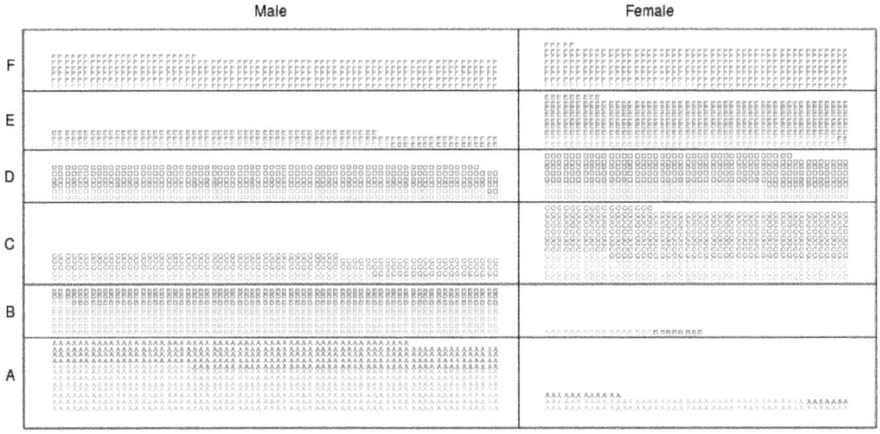

FIGURE 10.4
Simpon's Paradox illustrated. Admitted students are in green. Students are now stratified by the major to which they applied.

Average after stratifying In this plot, we can see that if we condition or stratify by major, and then look at differences, we control for the confounder and this effect goes away.

```
y=cbind(admissions[1:6,3],admissions[7:12,3])
matplot(1:6,y,xaxt="n",xlab="major",ylab="percent",col=c("blue","red"),cex=1.5)
```

```
axis(1,1:6,LETTERS[1:6])
legend("topright",c("Male","Female"),col=c("blue","red"),pch=c("1","2"),
       box.lty=0)
```

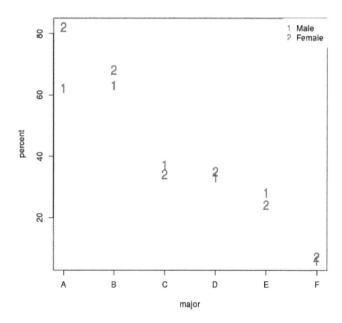

FIGURE 10.5
Admission percentage by major for each gender.

The average difference by major is actually 3.5% higher for women.

```
mean(y[,1]-y[,2])
```

```
## [1] -3.5
```

Simpson's Paradox in baseball Simpson's Paradox is commonly seen in baseball statistics. Here is a well known example in which David Justice had a higher batting average than Derek Jeter in both 1995 and 1996, but Jeter had a higher overall average:

	1995	1996	Combined
Derek Jeter	12/48 (.250)	183/582 (.314)	195/630 (.310)
David Justice	104/411 (.253)	45/140 (.321)	149/551 (.270)

The confounder here is games played. Jeter played more games during the year he batted better, while the opposite is true for Justice.

10.2   Confounding: High-Throughput Example

To describe the problem of confounding with a real example, we will use a dataset from this paper[2] that claimed that roughly 50% of genes where differentially expressed when comparing blood from two ethnic groups. We include the data in one of our data packages:

```
library(Biobase) ##available from Bioconductor
library(genefilter)
library(GSE5859) ##available from github
data(GSE5859)
```

We can extract the gene expression data and sample information table using the Bioconductor functions exprs and pData like this:

```
geneExpression = exprs(e)
sampleInfo = pData(e)
```

Note that some samples were processed at different times.

```
head(sampleInfo$date)
```

```
## [1] "2003-02-04" "2003-02-04" "2002-12-17" "2003-01-30" "2003-01-03"
## [6] "2003-01-16"
```

This is an extraneous variable and should not affect the values in geneExpression. However, as we have seen in previous analyses, it does appear to have an effect. We will therefore explore this here.

We can immediately see that year and ethnicity are almost completely confounded:

```
year = factor( format(sampleInfo$date,"%y") )
tab = table(year,sampleInfo$ethnicity)
print(tab)
```

```
##
## year ASN CEU HAN
##   02   0  32   0
##   03   0  54   0
##   04   0  13   0
##   05  80   3   0
##   06   2   0  24
```

By running a t-test and creating a volcano plot, we note that thousands of genes appear to be differentially expressed between ethnicities. Yet when we perform a similar comparison only on the CEU population between the years 2002 and 2003, we again obtain thousands of differentially expressed genes:

```
library(genefilter)
```

```
##remove control genes
out <- grep("AFFX",rownames(geneExpression))
```

---

[2]http://www.ncbi.nlm.nih.gov/pubmed/17206142

```
eth <- sampleInfo$ethnicity
ind<- which(eth%in%c("CEU","ASN"))
res1 <- rowttests(geneExpression[-out,ind],droplevels(eth[ind]))
ind <- which(year%in%c("02","03") & eth=="CEU")
res2 <- rowttests(geneExpression[-out,ind],droplevels(year[ind]))

XLIM <- max(abs(c(res1$dm,res2$dm)))*c(-1,1)
YLIM <- range(-log10(c(res1$p,res2$p)))
mypar(1,2)
plot(res1$dm,-log10(res1$p),xlim=XLIM,ylim=YLIM,
    xlab="Effect size",ylab="-log10(p-value)",main="Populations")
plot(res2$dm,-log10(res2$p),xlim=XLIM,ylim=YLIM,
    xlab="Effect size",ylab="-log10(p-value)",main="2003 v 2002")
```

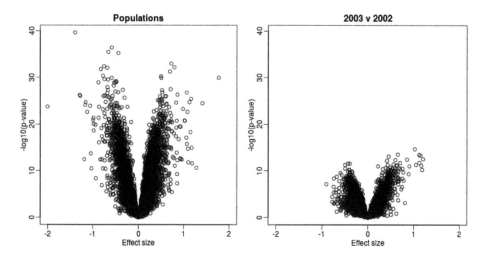

FIGURE 10.6
Volcano plots for gene expression data. Comparison by ethnicity (left) and by year within one ethnicity (right).

10.3   Exercises

Load the admissions data from the `dagdata` package (which is available from the genomicsclass repository):

```
library(dagdata)
data(admissions)
```

Familiarize yourself with this table:

admissions

1. Let's compute the proportion of men who were accepted:

```
index = which(admissions$Gender==1)
accepted= sum(admissions$Number[index] * admissions$Percent[index]/100)
applied = sum(admissions$Number[index])
accepted/applied
```

What is the proportion of women that were accepted?

2. Now that we have observed different acceptance rates between genders, test for the significance of this result.

If you perform an independence test, what is the p-value?

Now notice that looking at the data by major, the differences disappear.

How can this be? This is referred to as Simpson's Paradox. In the following questions we will try to decipher why this is happening.

3. We can quantify how "hard" a major is by using the percent of students that were accepted. Compute the percent that were accepted (regardless of gender) to each major and call this vector H.

Which is the hardest major?

4. What proportion is accepted for this major?

5. For men, what is the correlation between the number of applications across majors and H?

6. For women, what is the correlation between the number of applications across majors and H?

7. Given the answers to the above, which best explains the differences in admission percentages when we combine majors?

    (a) We made a coding mistake when computing the overall admissions percentages.

    (b) There were more total number of women applications which made the denominator much bigger.

    (c) There is confounding between gender and preference for "hard" majors: females are more likely to apply to harder majors.

    (d) The sample size for the individual majors was not large enough to draw the correct conclusion.

10.4 Discovering Batch Effects with EDA

Now that we understand PCA, we are going to demonstrate how we use it in practice with an emphasis on exploratory data analysis. To illustrate, we will go through an actual dataset that has not been sanitized for teaching purposes. We start with the raw data as it was provided in the public repository. The only step we did for you is to preprocess these data and create an R package with a preformed Bioconductor object.

10.5   Gene Expression Data

Start by loading the data:

```
library(rafalib)
library(Biobase)
library(GSE5859) ##Available from GitHub
data(GSE5859)
```

We start by exploring the sample correlation matrix and noting that one pair has a correlation of 1. This must mean that the same sample was uploaded twice to the public repository, but given different names. The following code identifies this sample and removes it.

```
cors <- cor(exprs(e))
Pairs=which(abs(cors)>0.9999,arr.ind=TRUE)
out = Pairs[which(Pairs[,1]<Pairs[,2]),,drop=FALSE]
if(length(out[,2])>0) e=e[,-out[2]]
```

We also remove control probes from the analysis:

```
out <- grep("AFFX",featureNames(e))
e <- e[-out,]
```

Now we are ready to proceed. We will create a detrended gene expression data matrix and extract the dates and outcome of interest from the sample annotation table.

```
y <- exprs(e)-rowMeans(exprs(e))
dates <- pData(e)$date
eth <- pData(e)$ethnicity
```

The original dataset did not include sex in the sample information. We did this for you in the subset dataset we provided for illustrative purposes. In the code below, we show how we predict the sex of each sample. The basic idea is to look at the median gene expression levels on Y chromosome genes. Males should have much higher values. To do this, we need to upload an annotation package that provides information for the features of the platform used in this experiment:

```
annotation(e)
```

```
## [1] "hgfocus"
```

We need to download and install the `hgfocus.db` package and then extract the chromosome location information.

```
library(hgfocus.db) ##install from Bioconductor
map2gene <- mapIds(hgfocus.db, keys=featureNames(e),
                column="ENTREZID", keytype="PROBEID",
                multiVals="first")
library(Homo.sapiens)
map2chr <- mapIds(Homo.sapiens, keys=map2gene,
                column="TXCHROM", keytype="ENTREZID",
                multiVals="first")
chryexp <- colMeans(y[which(unlist(map2chr)=="chrY"),])
```

If we create a histogram of the median gene expression values on chromosome Y, we clearly see two modes which must be females and males:

```
mypar()
hist(chryexp)
```

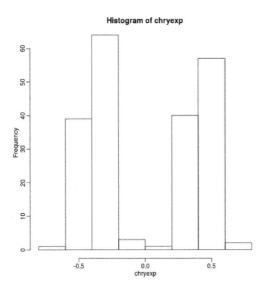

FIGURE 10.7
Histogram of median expression y-axis. We can see females and males.

So we can predict sex this way:

```
sex <- factor(ifelse(chryexp<0,"F","M"))
```

Calculating the PCs We have shown how we can compute principal components using:

```
s <- svd(y)
dim(s$v)
```

```
## [1] 207 207
```

We can also use `prcomp` which creates an object with just the PCs and also demeans by default. They provide practically the same principal components so we continue the analysis with the object *s*.

Variance explained A first step in determining how much sample correlation induced *structure* there is in the data.

```
library(RColorBrewer)
cols=colorRampPalette(rev(brewer.pal(11,"RdBu")))(100)
n <- ncol(y)
image(1:n,1:n,cor(y),xlab="samples",ylab="samples",col=cols,zlim=c(-1,1))
```

Here we are using the term *structure* to refer to the deviation from what one would see if the samples were in fact independent from each other. The plot above clearly shows groups of samples that are more correlated between themselves than to others.

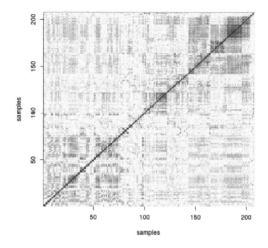

FIGURE 10.8
Image of correlations. Cell (i,j) represents correlation between samples i and j. Red is high, white is 0 and red is negative.

One simple exploratory plot we make to determine how many principal components we need to describe this *structure* is the variance-explained plot. This is what the variance explained for the PCs would look like if data were independent :

```
y0 <- matrix( rnorm( nrow(y)*ncol(y) ) , nrow(y), ncol(y) )
d0 <- svd(y0)$d
plot(d0^2/sum(d0^2),ylim=c(0,.25))
```

Instead we see this:

```
plot(s$d^2/sum(s$d^2))
```

At least 20 or so PCs appear to be higher than what we would expect with independent data. A next step is to try to explain these PCs with measured variables. Is this driven by ethnicity? Sex? Date? Or something else?

MDS plot As previously shown, we can make MDS plots to start exploring the data to answer these questions. One way to explore the relationship between variables of interest and PCs is to use color to denote these variables. For example, here are the first two PCs with color representing ethnicity:

```
cols = as.numeric(eth)
mypar()
plot(s$v[,1],s$v[,2],col=cols,pch=16,
     xlab="PC1",ylab="PC2")
legend("bottomleft",levels(eth),col=seq(along=levels(eth)),pch=16)
```

There is a very clear association between the first PC and ethnicity. However, we also see that for the orange points there are sub-clusters. We know from previous analyses that ethnicity and preprocessing date are correlated:

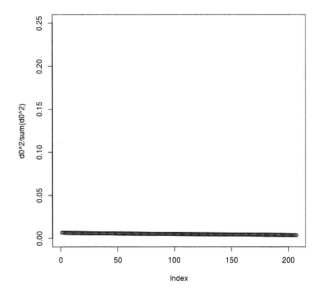

FIGURE 10.9
Variance explained plot for simulated independent data.

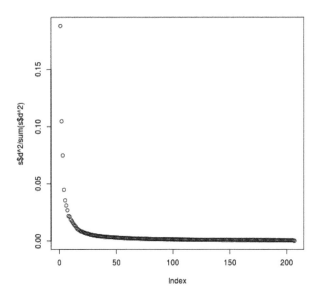

FIGURE 10.10
Variance explained plot for gene expression data.

```
year = factor(format(dates,"%y"))
table(year,eth)
```

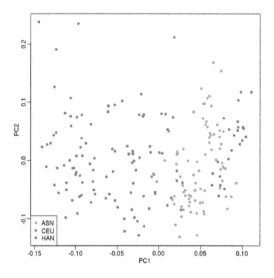

FIGURE 10.11
First two PCs for gene expression data with color representing ethnicity.

```
##       eth
## year ASN CEU HAN
##   02   0  32   0
##   03   0  54   0
##   04   0  13   0
##   05  80   3   0
##   06   2   0  23
```

So explore the possibility of date being a major source of variability by looking at the same plot, but now with color representing year:

```
cols = as.numeric(year)
mypar()
plot(s$v[,1],s$v[,2],col=cols,pch=16,
    xlab="PC1",ylab="PC2")
legend("bottomleft",levels(year),col=seq(along=levels(year)),pch=16)
```

We see that year is also very correlated with the first PC. So which variable is driving this? Given the high level of confounding, it is not easy to parse out. Nonetheless, in the assessment questions and below, we provide some further exploratory approaches.

Boxplot of PCs The structure seen in the plot of the between sample correlations shows a complex structure that seems to have more than 5 factors (one for each year). It certainly has more complexity than what would be explained by ethnicity. We can also explore the correlation with months.

```
month <- format(dates,"%y%m")
length( unique(month))
```

```
## [1] 21
```

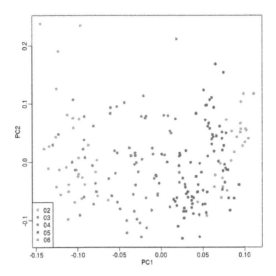

FIGURE 10.12
First two PCs for gene expression data with color representing processing year.

Because there are so many months (21), it becomes complicated to use color. Instead we can stratify by month and look at boxplots of our PCs:

```
variable <- as.numeric(month)
mypar(2,2)
for(i in 1:4){
  boxplot(split(s$v[,i],variable),las=2,range=0)
  stripchart(split(s$v[,i],variable),add=TRUE,vertical=TRUE,pch=1,cex=.5,col=1)
  }
```

Here we see that month has a very strong correlation with the first PC, even when stratifying by ethnic group as well as some of the others. Remember that samples processed between 2002-2004 are all from the same ethnic group. In cases such as these, in which we have many samples, we can use an analysis of variance to see which PCs correlate with month:

```
corr <- sapply(1:ncol(s$v),function(i){
  fit <- lm(s$v[,i]~as.factor(month))
  return( summary(fit)$adj.r.squared  )
  })
mypar()
plot(seq(along=corr), corr, xlab="PC")
```

We see a very strong correlation with the first PC and relatively strong correlations for the first 20 or so PCs. We can also compute F-statistics comparing within month to across month variability:

```
Fstats<- sapply(1:ncol(s$v),function(i){
   fit <- lm(s$v[,i]~as.factor(month))
   Fstat <- summary(aov(fit))[[1]][1,4]
   return(Fstat)
   })
```

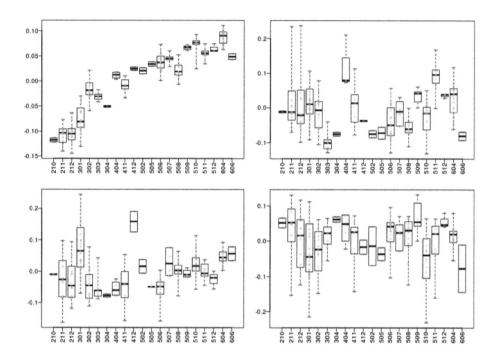

FIGURE 10.13
Boxplot of first four PCs stratified by month.

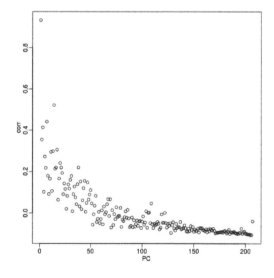

FIGURE 10.14
Adjusted R-squared after fitting a model with each month as a factor to each PC.

```
mypar()
plot(seq(along=Fstats),sqrt(Fstats))
p <- length(unique(month))
abline(h=sqrt(qf(0.995,p-1,ncol(s$v)-1)))
```

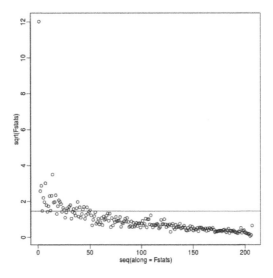

FIGURE 10.15
Square root of F-statistics from an analysis of variance to explain PCs with month.

We have seen how PCA combined with EDA can be a powerful technique to detect and understand batches. In a later section, we will see how we can use the PCs as estimates in factor analysis to improve model estimates.

10.6   Exercises

We will use the Bioconductor package `Biobase` which you can install with `install_bioc` function from `rafalib`:

Load the data for this gene expression dataset:

```
library(Biobase)
library(GSE5859)
data(GSE5859)
```

This is the original dataset from which we selected the subset used in `GSE5859Subset`.
We can extract the gene expression data and sample information table using the Bioconductor functions `exprs` and `pData` like this:

```
geneExpression = exprs(e)
sampleInfo = pData(e)
```

1. Familiarize yourself with the `sampleInfo` table. Note that some samples were processed at different times. This is an extraneous variable and should not affect

the values in `geneExpression`. However, as we have seen in previous analyses, it does appear to have an effect so we will explore this here.

You can extract the year from each date like this:

```
year = format(sampleInfo$date,"%y")
```

Note that ethnic group and year is almost perfectly confounded:

```
table(year,sampleInfo$ethnicity)
```

1. For how many of these years do we have more than one ethnicity represented?

2. Repeat the above exercise, but now, instead of year, consider the month as well. Specifically, instead of the `year` variable defined above use:

```
month.year = format(sampleInfo$date,"%m%y")
```

For what **proportion** of these `month.year` values do we have more than one ethnicity represented?

3. Perform a t-test (use `rowttests`) comparing CEU samples processed in 2002 to those processed in 2003. Then use the `qvalue` package to obtain q-values for each gene.

How many genes have q-values < 0.05 ?

4. What is the estimate of `pi0` provided by `qvalue`:

5. Now perform a t-test (use `rowttests`) comparing CEU samples processed in 2003 to those processed in 2004. Then use the `qvalue` package to obtain q-values for each gene. How many genes have q-values less than 0.05?

6. Now we are going to compare ethnicities as was done in the original publication in which these data were first presented. Use the `qvalue` function to compare the ASN population to the CEU population. Once again, use the `qvalue` function to obtain q-values.

How many genes have q-values < 0.05 ?

7. Over 80% of genes are called differentially expressed between ethnic groups. However, due to the confounding with processing date, we need to confirm these differences are actually due to ethnicity. This will not be easy due to the almost perfect confounding. However, above we noted that two groups were represented in 2005. Just like we stratified by majors to remove the "major effect" in our admissions example, here we can stratify by year and perform a t-test comparing ASN and CEU, but only for samples processed in 2005.

How many genes have q-values < 0.05 ?

Notice the dramatic drop in the number of genes with q-value < 0.05 when we fix the year. However, the sample size is much smaller in this latest analysis which means we have less power:

```
table(sampleInfo$ethnicity[index])
```

8. To provide a more balanced comparison, we repeat the analysis, but now taking 3 random CEU samples from 2002. Repeat the analysis above, but comparing the ASN from 2005 to three random CEU samples from 2002. Set the seed at 3, `set.seed(3)`

How many genes have q-values < 0.05 ?

10.7   Motivation for Statistical Approaches

Data example   To illustrate how we can adjust for batch effects using statistical methods, we will create a data example in which the outcome of interest is somewhat confounded with batch, but not completely. To aid with the illustration and assessment of the methods we demonstrate, we will also select an outcome for which we have an expectation of what genes should be differentially expressed. Namely, we make sex the outcome of interest and expect genes on the Y chromosome to be differentially expressed. We may also see genes from the X chromosome as differentially expressed since some escape X inactivation. The data with these properties is the one included in this dataset:

```
##available from course github repository
library(GSE5859Subset)
data(GSE5859Subset)
```

We can see the correlation between sex and month:

```
month <- format(sampleInfo$date,"%m")
table(sampleInfo$group, month)
```

```
##     month
##      06 10
##   0  9  3
##   1  3  9
```

To illustrate the confounding, we will pick some genes to show in a heatmap plot. We pick 1) all Y chromosome genes, 2) some genes that we see correlate with batch, and 3) some randomly selected genes. The image below (code not shown) shows high values in red, low values in blue, middle values in yellow. Each column is a sample and each row is one of the randomly selected genes:

In the plot above, the first 12 columns are females (1s) and the last 12 columns are males (0s). We can see some Y chromosome genes towards the top since they are blue for females and red from males. We can also see some genes that correlate with month towards the bottom of the image. Some genes are low in June (6) and high in October (10), while others do the opposite. The month effect is not as clear as the sex effect, but it is certainly present.

In what follows, we will imitate the typical analysis we would do in practice. We will act as if we don't know which genes are supposed to be differentially expressed between males and females, find genes that are differentially expressed, and the evaluate these methods by comparing to what we expect to be correct. Note while in the plot we only show a few genes, for the analysis we analyze all 8,793.

Assessment plots and summaries   For the assessment of the methods we present, we will assume that autosomal (not on chromosome X or Y) genes on the list are likely false positives. We will also assume that genes on chromosome Y are likely true positives. Chromosome X genes could go either way. This gives us the opportunity to estimate both specificity and sensitivity. Since in practice we rarely know the "truth", these evaluations are not possible. Simulations are therefore commonly used for evaluation purposes: we know the truth because we construct the data. However, simulations are at risk of not capturing all the nuances of real experimental data. In contrast, this dataset is an experimental dataset.

In the next sections, we will use the histogram p-values to evaluate the specificity (low

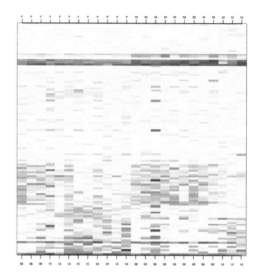

FIGURE 10.16
Image of gene expression data for genes selected to show difference in group as well as the batch effect, along with some randomly chosen genes.

false positive rates) of the batch adjustment procedures presented here. Because the autosomal genes are not expected to be differentially expressed, we should see a a flat p-value histogram. To evaluate sensitivity (low false negative rates), we will report the number of the reported genes on chromosome X and chromosome Y for which we reject the null hypothesis. We also include a volcano plot with a horizontal dashed line separating the genes called significant from those that are not, and colors used to highlight chromosome X and Y genes.

Below are the results of applying a naive t-test and report genes with q-values smaller than 0.1.

```
library(qvalue)
res <- rowttests(geneExpression,as.factor( sampleInfo$group ))
mypar(1,2)
hist(res$p.value[which(!chr%in%c("chrX","chrY") )],main="",ylim=c(0,1300))

plot(res$dm,-log10(res$p.value))
points(res$dm[which(chr=="chrX")],-log10(res$p.value[which(chr=="chrX")]),col=1,pch=16)
points(res$dm[which(chr=="chrY")],-log10(res$p.value[which(chr=="chrY")]),col=2,pch=16,
       xlab="Effect size",ylab="-log10(p-value)")
legend("bottomright",c("chrX","chrY"),col=1:2,pch=16)
qvals <- qvalue(res$p.value)$qvalue
index <- which(qvals<0.1)
abline(h=-log10(max(res$p.value[index])))

cat("Total genes with q-value < 0.1: ",length(index),"\n",
    "Number of selected genes on chrY: ", sum(chr[index]=="chrY",na.rm=TRUE),"\n",
    "Number of selected genes on chrX: ", sum(chr[index]=="chrX",na.rm=TRUE),sep="")
```

```
## Total genes with q-value < 0.1: 59
## Number of selected genes on chrY: 8
## Number of selected genes on chrX: 12
```

We immediately note that the histogram is not flat. Instead, low p-values are over-

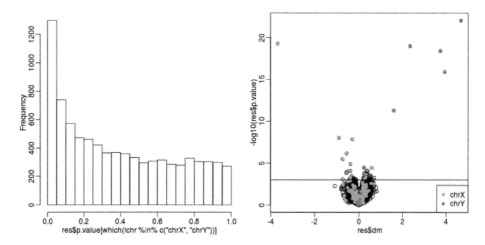

FIGURE 10.17
p-value histogram and volcano plot for comparison between sexes. The Y chromosome genes (considered to be positives) are highlighted in red. The X chromosome genes (a subset is considered to be positive) are shown in green.

represented. Furthermore, more than half of the genes on the final list are autosomal. We now describe two statistical solutions and try to improve on this.

10.8 Adjusting for Batch Effects with Linear Models

We have already observed that processing date has an effect on gene expression. We will therefore try to *adjust* for this by including it in a model. When we perform a t-test comparing the two groups, it is equivalent to fitting the following linear model:

$$Y_{ij} = \alpha_j + x_i\beta_j + \varepsilon_{ij}$$

to each gene j with $x_i = 1$ if subject i is female and 0 otherwise. Note that β_j represents the estimated difference for gene j and ε_{ij} represents the within group variation. So what is the problem?

The theory we described in the linear models chapter assumes that the error terms are independent. We know that this is not the case for all genes because we know the error terms from October will be more alike to each other than the June error terms. We can *adjust* for this by including a term that models this effect:

$$Y_{ij} = \alpha_j + x_i\beta_j + z_i\gamma_j + \varepsilon_{ij}.$$

Here $z_i = 1$ if sample i was processed in October and 0 otherwise and γ_j is the month effect for gene j. This an example of how linear models give us much more flexibility than procedures such as the t-test.

We construct a model matrix that includes batch.

```
sex <- sampleInfo$group
X <- model.matrix(~sex+batch)
```

Now we can fit a model for each gene. For example, note the difference between the original model and one that has been adjusted for batch:

```
j <- 7635
y <- geneExpression[j,]
X0 <- model.matrix(~sex)
fit <- lm(y~X0-1)
summary(fit)$coef
```

```
##                  Estimate Std. Error   t value     Pr(>|t|)
## X0(Intercept)  6.9555747  0.2166035 32.112008 5.611901e-20
## X0sex         -0.6556865  0.3063237 -2.140502 4.365102e-02
```

```
X <- model.matrix(~sex+batch)
fit <- lm(y~X)
summary(fit)$coef
```

```
##                Estimate Std. Error    t value     Pr(>|t|)
## (Intercept)  7.26329968  0.1605560 45.2384140 2.036006e-22
## Xsex        -0.04023663  0.2427379 -0.1657616 8.699300e-01
## Xbatch10    -1.23089977  0.2427379 -5.0709009 5.070727e-05
```

We then fit this new model for each gene. For instance, we can use `sapply` to recover the estimated coefficient and p-value in the following way:

```
res <- t( sapply(1:nrow(geneExpression),function(j){
  y <- geneExpression[j,]
  fit <- lm(y~X-1)
  summary(fit)$coef[2,c(1,4)]
} ) )


##turn into data.frame so we can use the same code for plots as above
res <- data.frame(res)
names(res) <- c("dm","p.value")

mypar(1,2)
hist(res$p.value[which(!chr%in%c("chrX","chrY") )],main="",ylim=c(0,1300))

plot(res$dm,-log10(res$p.value))
points(res$dm[which(chr=="chrX")],-log10(res$p.value[which(chr=="chrX")]),col=1,pch=16)
points(res$dm[which(chr=="chrY")],-log10(res$p.value[which(chr=="chrY")]),col=2,pch=16,
       xlab="Effect size",ylab="-log10(p-value)")
legend("bottomright",c("chrX","chrY"),col=1:2,pch=16)
qvals <- qvalue(res$p.value)$qvalue
index <- which(qvals<0.1)
abline(h=-log10(max(res$p.value[index])))

cat("Total genes with q-value < 0.1: ",length(index),"\n",
    "Number of selected genes on chrY: ", sum(chr[index]=="chrY",na.rm=TRUE),"\n",
    "Number of selected genes on chrX: ", sum(chr[index]=="chrX",na.rm=TRUE),sep="")
```

```
## Total genes with q-value < 0.1: 17
## Number of selected genes on chrY: 6
## Number of selected genes on chrX: 9
```

There is a great improvement in specificity (less false positives) without much loss in sensitivity (we still find many chromosome Y genes). However, we still see some bias in the histogram. In a later section we will see that month does not perfectly account for the batch effect and that better estimates are possible.

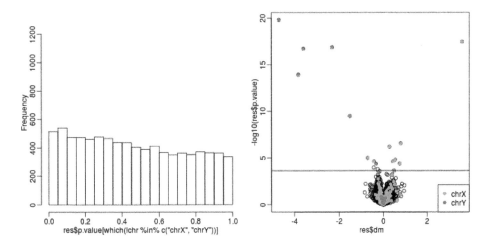

FIGURE 10.18
p-value histogram and volcano plot for comparison between sexes after adjustment for month. The Y chromosome genes (considered to be positives) are highlighted in red. The X chromosome genes (a subset is considered to be positive) are shown in green.

A note on computing efficiency In the code above, the design matrix does not change within the iterations we are computing $(X^\top X)^{-1}$ repeatedly and applying to each gene. Instead we can perform this calculation in one matrix algebra calculation by computing it once and then obtaining all the betas by multiplying $(X^\top X)^{-1} X^\top Y$ with the columns of Y representing genes in this case. The `limma` package has an implementation of this idea (using the QR decomposition). Notice how much faster this is:

```
library(limma)
X <- model.matrix(~sex+batch)
fit <- lmFit(geneExpression,X)
```

The estimated regression coefficients for each gene are obtained like this:

```
dim( fit$coef)
```

```
## [1] 8793    3
```

We have one estimate for each gene. To obtain p-values for one of these, we have to construct the ratios:

```
k <- 2 ##second coef
ses <- fit$stdev.unscaled[,k]*fit$sigma
ttest <- fit$coef[,k]/ses
pvals <- 2*pt(-abs(ttest),fit$df)
```

Combat Combat[3] is a popular method and is based on using linear models to adjust for batch effects. It fits a hierarchical model to estimate and remove row specific batch effects. Combat uses a modular approach. In a first step, what is considered to be a batch effect is removed:

---

[3]http://biostatistics.oxfordjournals.org/content/8/1/118.short

```
library(sva) #available from Bioconductor
mod <- model.matrix(~sex)
cleandat <- ComBat(geneExpression,batch,mod)
```

```
## Found 2 batches
## Adjusting for 1 covariate(s) or covariate level(s)
## Standardizing Data across genes
## Fitting L/S model and finding priors
## Finding parametric adjustments
## Adjusting the Data
```

Then the results can be used to fit a model with our variable of interest:

```
res<-genefilter::rowttests(cleandat,factor(sex))
```

In this case, the results are less specific than what we obtain by fitting the simple linear model:

```
mypar(1,2)
hist(res$p.value[which(!chr%in%c("chrX","chrY") )],main="",ylim=c(0,1300))

plot(res$dm,-log10(res$p.value))
points(res$dm[which(chr=="chrX")],-log10(res$p.value[which(chr=="chrX")]),col=1,pch=16)
points(res$dm[which(chr=="chrY")],-log10(res$p.value[which(chr=="chrY")]),col=2,pch=16,
      xlab="Effect size",ylab="-log10(p-value)")
legend("bottomright",c("chrX","chrY"),col=1:2,pch=16)
qvals <- qvalue(res$p.value)$qvalue
index <- which(qvals<0.1)
abline(h=-log10(max(res$p.value[index])))
```

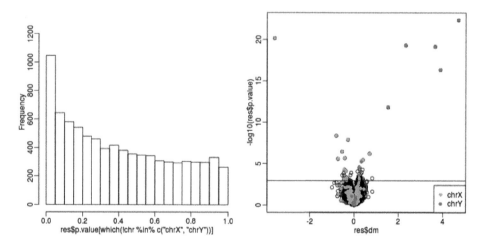

FIGURE 10.19
p-value histogram and volcano plot for comparison between sexes for Combat. The Y chromosome genes (considered to be positives) are highlighted in red. The X chromosome genes (a subset is considered to be positive) are shown in green.

```
cat("Total genes with q-value < 0.1: ",length(index),"\n",
    "Number of selected genes on chrY: ", sum(chr[index]=="chrY",na.rm=TRUE),"\n",
    "Number of selected genes on chrX: ", sum(chr[index]=="chrX",na.rm=TRUE),sep="")
```

```
## Total genes with q-value < 0.1: 68
## Number of selected genes on chrY: 8
## Number of selected genes on chrX: 16
```

10.9   Exercises

For the dataset we have been working with, models do not help due to the almost perfect confounding. This is one reason we created the subset dataset:

```
library(GSE5859Subset)
data(GSE5859Subset)
```

Here we purposely confounded month and group (sex), but not completely:

```
sex = sampleInfo$group
month = factor( format(sampleInfo$date,"%m"))
table( sampleInfo$group, month)
```

1. Using the functions `rowttests` and `qvalue` compare the two groups. Because this is a smaller dataset which decreases our power, we will use the more lenient FDR cut-off of 10%.

How many gene have q-values less than 0.1?

2. Note that `sampleInfo$group` here presents males and females. Thus, we expect differences to be in on chrY and, for genes that escape inactivation, chrX. We do not expect many autosomal genes to be different between males and females. This gives us an opportunity to evaluate false and true positives with experimental data. For example, we evaluate results using the proportion genes of the list that are on chrX or chrY.

For the list calculated above, what proportion of this list is on chrX or chrY?

3. We can also check how many of the chromosomes X and Y genes we detected as different. How many are on Y?

4. Now for the autosomal genes (not on chrX and chrY) for which q-value < 0.1, perform a t-test comparing samples processed in June to those processed in October.

What proportion of these have p-values <0.05 ?

5. The above result shows that the great majority of the autosomal genes show differences due to processing data. This provides further evidence that confounding is resulting in false positives. So we are going to try to model the month effect to better estimate the sex effect. We are going to use a linear model:

Which of the following creates the appropriate design matrix?

- A) `X = model.matrix(~sex+ethnicity)`

- B) `X = cbind(sex,as.numeric(month))`

- C) It can't be done with one line.

- D) `X = model.matrix(~sex+month)`

6. Now use the X defined above, to fit a regression model using `lm` for each gene. You can obtain p-values for estimated parameters using `summary`. Here is an example

```
X = model.matrix(~sex+month)
i = 234
y = geneExpression[i,]
fit = lm(y~X)
summary(fit)$coef
```

How many of the q-values for the group comparison are now <0.1? Note the big drop from what we obtained without the correction.

7. With this new list, what proportion of these are chrX and chrY?

Notice the big improvement.

8. How many on Y or X?

9. Now from the linear model above, extract the p-values related to the coefficient representing the October versus June differences using the same linear model.

How many of the q-values for the month comparison are now <0.1? This approach is basically the approach implemented by Combat.

10.10   Factor Analysis

Before we introduce the next type of statistical method for batch effect correction, we introduce the statistical idea that motivates the main idea: Factor Analysis. Factor Analysis was first developed over a century ago. Karl Pearson noted that correlation between different subjects when the correlation was computed across students. To explain this, he posed a model having one factor that was common across subjects for each student that explained this correlation:

$$Y_ij = \alpha_i W_1 + \varepsilon_{ij}$$

with Y_{ij} the grade for individual i on subject j and α_i representing the ability of student i to obtain good grades.

In this example, W_1 is a constant. Here we will motivate factor analysis with a slightly more complicated situation that resembles the presence of batch effects. We generate a random $N \times 6$ matrix $\mathbf{Y}$ with representing grades in six different subjects for N different children. We generate the data in a way that subjects are correlated with some more than others:

Sample correlations Note that we observe high correlation across the six subjects:

```
round(cor(Y),2)
```

```
##         Math Science   CS  Eng Hist Classics
## Math     1.00    0.67 0.64 0.34 0.29     0.28
## Science  0.67    1.00 0.65 0.29 0.29     0.26
## CS       0.64    0.65 1.00 0.35 0.30     0.29
## Eng      0.34    0.29 0.35 1.00 0.71     0.72
## Hist     0.29    0.29 0.30 0.71 1.00     0.68
## Classics 0.28    0.26 0.29 0.72 0.68     1.00
```

A graphical look shows that the correlation suggests a grouping of the subjects into STEM and the humanities.

In the figure below, high correlations are red, no correlation is white and negative correlations are blue (code not shown).

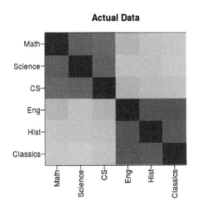
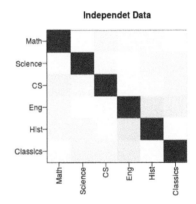

FIGURE 10.20
Images of correlation between columns. High correlation is red, no correlation is white, and negative correlation is blue.

The figure shows the following: there is correlation across all subjects, indicating that students have an underlying hidden factor (academic ability for example) that results in subjects begin correlated since students that test high in one subject tend to test high in the others. We also see that this correlation is higher with the STEM subjects and within the humanities subjects. This implies that there is probably another hidden factor that determines if students are better in STEM or humanities. We now show how these concepts can be explained with a statistical model.

Factor model Based on the plot above, we hypothesize that there are two hidden factors $\mathbf{W}_1$ and $\mathbf{W}_2$ and, to account for the observed correlation structure, we model the data in the following way:

$$Y_{ij} = \alpha_{i,1} W_{1,j} + \alpha_{i,2} W_{2,j} + \varepsilon_{ij}$$

The interpretation of these parameters are as follows: $\alpha_{i,1}$ is the overall academic ability for student i and $\alpha_{i,2}$ is the difference in ability between the STEM and humanities for student i. Now, can we estimate the W and α ?

Factor analysis and PCA It turns out that under certain assumptions, the first two principal components are optimal estimates for W_1 and W_2. So we can estimate them like this:

```
s <- svd(Y)
What <- t(s$v[,1:2])
colnames(What)<-colnames(Y)
round(What,2)
```

```
##        Math Science    CS  Eng Hist Classics
## [1,]   0.36    0.36  0.36 0.47 0.43     0.45
## [2,]  -0.44   -0.49 -0.42 0.34 0.34     0.39
```

As expected, the first factor is close to a constant and will help explain the observed correlation across all subjects, while the second is a factor differs between STEM and humanities. We can now use these estimates in the model:

$$Y_{ij} = \alpha_{i,1}\hat{W}_{1,j} + \alpha_{i,2}\hat{W}_{2,j} + \varepsilon_{ij}$$

and we can now fit the model and note that it explains a large percent of the variability.

```
fit = s$u[,1:2]%*% (s$d[1:2]*What)
var(as.vector(fit))/var(as.vector(Y))
```

```
## [1] 0.7880933
```

The important lesson here is that when we have correlated units, the standard linear models are not appropriate. We need to account for the observed structure somehow. Factor analysis is a powerful way of achieving this.

Factor analysis in general In high-throughput data, it is quite common to see correlation structure. For example, notice the complex correlations we see across samples in the plot below. These are the correlations for a gene expression experiment with columns ordered by date:

```
library(Biobase)
library(GSE5859)
data(GSE5859)
n <- nrow(pData(e))
o <- order(pData(e)$date)
Y=exprs(e)[,o]
cors=cor(Y-rowMeans(Y))
cols=colorRampPalette(rev(brewer.pal(11,"RdBu")))(100)
```

```
mypar()
image(1:n,1:n,cors,col=cols,xlab="samples",ylab="samples",zlim=c(-1,1))
```

Two factors will not be enough to model the observed correlation structure. However, a more general factor model can be useful:

$$Y_{ij} = \sum_{k=1}^{K} \alpha_{i,k}W_{j,k} + \varepsilon_{ij}$$

And we can use PCA to estimate $\mathbf{W}_1, \ldots, \mathbf{W}_K$. However, choosing k is a challenge and a topic of current research. In the next section we describe how exploratory data analysis might help.

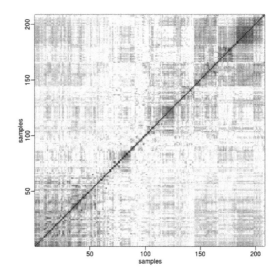

FIGURE 10.21

Image of correlations. Cell (i,j) represents correlation between samples i and j. Red is high, white is 0 and red is negative.

10.11 Exercises

We will continue to use this dataset:

```
library(Biobase)
library(GSE5859Subset)
data(GSE5859Subset)
```

1. Suppose you want to make an MA-plot of the first two samples y =
   geneExpression[,1:2]. Which of the following projections gives us the projection of y so that column 2 versus column 1 is an MA plot?

   A. $y \begin{pmatrix} 1/\sqrt{2} & 1/\sqrt{2} \\ 1\sqrt{2} & -1/\sqrt{2} \end{pmatrix}$ B. $y \begin{pmatrix} 1 & 1 \\ 1 & -1 \end{pmatrix}$ C. $\begin{pmatrix} 1 & 1 \\ 1 & -1 \end{pmatrix} y$ D. $\begin{pmatrix} 1 & 1 \\ 1 & -1 \end{pmatrix} y^{\top}$

2. Say Y is $M \times N$, in the SVD $Y = UDV^{\top}$ which of the following is not correct?

- A) $DV^{\top}$ are the new coordinates for the projection $U^{\top}Y$

- B) UD are the new coordinates for the projection YV

- C) D are the coordinates of the projection $U^{\top}Y$

- D) $U^{\top}Y$ is a projection from an N-dimensional to M-dimensional subspace.

3. Define:

```
y = geneExpression - rowMeans(geneExpression)
```

Compute and plot an image of the correlation for each sample. Make two image plots of these correlations. In the first one, plot the correlation as image. In the second, order the samples by date and then plot an image of the correlation. The only difference in these plots is the order in which the samples are plotted.

Based on these plots, which of the following would you say is true?

- A) The samples appear to be completely independent of each other.

- B) Sex seems to be creating structures as evidenced by the two clusters of highly correlated samples.

- C) There appear to be only two factors completely driven by month.

- D) The fact that in the plot ordered by month we see two groups mainly driven by month, and within these we see subgroups driven by date, seems to suggest date more than month per se are the hidden factors.

4. Based on the correlation plots above, we could argue that there are at least two hidden factors. Using PCA estimate these two factors. Specifically, apply the `svd` to `y` and use the first two PCs as estimates.

Which command gives us these estimates?

- A) `pcs = svd(y)$v[1:2,]`

- B) `pcs = svd(y)$v[,1:2]`

- C) `pcs = svd(y)$u[,1:2]`

- D) `pcs = svd(y)$d[1:2]`

5. Plot each of the estimated factors ordered by date. Use color to denote month. The first factor is clearly related to date. Which of the following appear to be most different according to this factor?

- A) June 23 and June 27

- B) Oct 07 and Oct 28

- C) June 10 and June 23

- D) June 15 and June 24

6. Use the `svd` function to obtain the principal components (PCs) for our detrended gene expression data `y`.

How many PCs explain more than 10% of the variability?

7. Which PC most correlates (negative or positive correlation) with month?

8. What is this correlation (in absolute value)?

9. Which PC most correlates (negative or positive correlation) with sex?

10. What is this correlation (in absolute value)?

11. Now instead of using month, which we have shown does not quite describe the batch, add the two estimated factors `s$v[,1:2]` to the linear model we used above. Apply this model to each gene and compute q-values for the sex difference. How many q-values < 0.1 for the sex comparison?

12. What proportion of the genes are on chromosomes X and Y?

10.12 Modeling Batch Effects with Factor Analysis

We continue to use this data set:

```
library(GSE5859Subset)
data(GSE5859Subset)
```

Below is the image we showed earlier with a subset of genes showing both the sex effect and the month time effects, but now with an image showing the sample to sample correlations (computed on all genes) showing the complex structure of the data (code not shown):

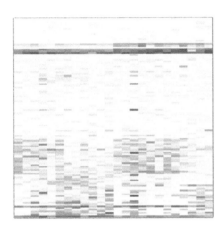 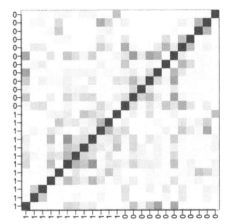

FIGURE 10.22
Image of subset gene expression data (left) and image of correlations for this dataset (right).

We have seen how the approach that assumes month explains the batch and adjusts with linear models perform relatively well. However, there was still room for improvement. This is most likely due to the fact that month is only a surrogate for some hidden factor or factors that actually induces structure or between sample correlation.

What is a batch? Here is a plot of dates for each sample, with color representing month:

```
times <-sampleInfo$date
mypar(1,1)
o=order(times)
plot(times[o],pch=21,bg=as.numeric(batch)[o],ylab="Date")
```

We note that there is more than one day per month. Could day have an effect as well? We can use PCA and EDA to try to answer this question. Here is a plot of the first principal component ordered by date:

```
s <- svd(y)
mypar(1,1)
o<-order(times)
cols <- as.numeric( batch)
```

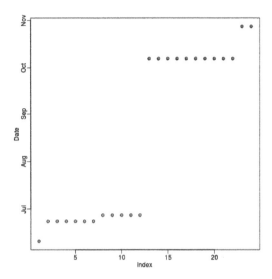

FIGURE 10.23
Dates with color denoting month.

```
plot(s$v[o,1],pch=21,cex=1.25,bg=cols[o],ylab="First PC",xlab="Date order")
legend("topleft",c("Month 1","Month 2"),col=1:2,pch=16,box.lwd=0)
```

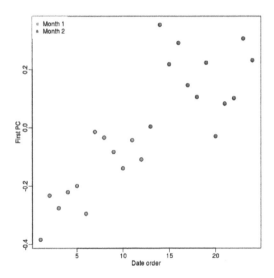

FIGURE 10.24
First PC plotted against order by date with colors representing month.

Day seems to be highly correlated with the first PC, which explains a high percentage
of the variability:

```
mypar(1,1)
plot(s$d^2/sum(s$d^2),ylab="% variance explained",xlab="Principal component")
```

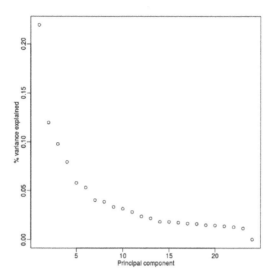

FIGURE 10.25
Variance explained.

Further exploration shows that the first six or so PC seem to be at least partially driven by date:

```
mypar(3,4)
for(i in 1:12){
  days <- gsub("2005-","",times)
  boxplot(split(s$v[,i],gsub("2005-","",days)))
}
```

So what happens if we simply remove the top six PC from the data and then perform a t-test?

```
D <- s$d; D[1:4]<-0 #take out first 2
cleandat <- sweep(s$u,2,D,"*")%*%t(s$v)
res <-rowttests(cleandat,factor(sex))
```

This does remove the batch effect, but it seems we have also removed much of the biological effect we are interested in. In fact, no genes have q-value <0.1 anymore.

```
library(qvalue)
mypar(1,2)
hist(res$p.value[which(!chr%in%c("chrX","chrY") )],main="",ylim=c(0,1300))

plot(res$dm,-log10(res$p.value))
points(res$dm[which(chr=="chrX")],-log10(res$p.value[which(chr=="chrX")]),col=1,pch=16)
points(res$dm[which(chr=="chrY")],-log10(res$p.value[which(chr=="chrY")]),col=2,pch=16,
       xlab="Effect size",ylab="-log10(p-value)")
legend("bottomright",c("chrX","chrY"),col=1:2,pch=16)

qvals <- qvalue(res$p.value)$qvalue
index <- which(qvals<0.1)

cat("Total genes with q-value < 0.1: ",length(index),"\n",
    "Number of selected genes on chrY: ", sum(chr[index]=="chrY",na.rm=TRUE),"\n",
    "Number of selected genes on chrX: ", sum(chr[index]=="chrX",na.rm=TRUE),sep="")
```

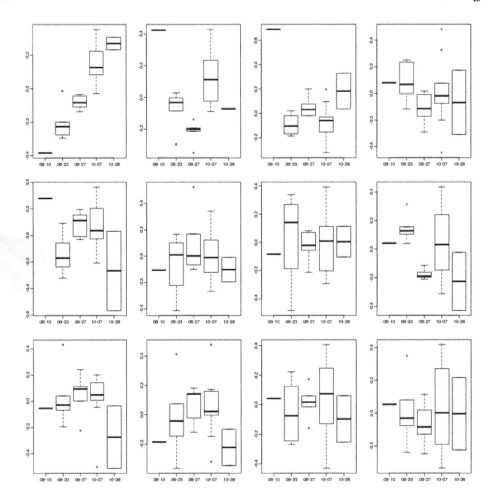

FIGURE 10.26
First 12 PCs stratified by dates.

```
## Total genes with q-value < 0.1: 0
## Number of selected genes on chrY: 0
## Number of selected genes on chrX: 0
```

In this case we seem to have over corrected since we now recover many fewer chromosome Y genes and the p-value histogram shows a dearth of small p-values that makes the distribution non-uniform. Because sex is probably correlated with some of the first PCs, this may be a case of "throwing out the baby with the bath water".

Surrogate Variable Analysis A solution to the problem of over-correcting and removing the variability associated with the outcome of interest is fit models with both the covariate of interest, as well as those believed to be batches. An example of an approach that does this is Surrogate Variable Analysis[4] (SVA).

The basic idea of SVA is to first estimate the factors, but taking care not to include the outcome of interest. To do this, an interactive approach is used in which each row is

---

[4]http://www.ncbi.nlm.nih.gov/pmc/articles/PMC1994707/

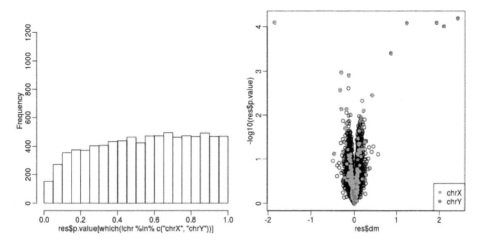

FIGURE 10.27
p-value histogram and volcano plot after blindly removing the first two PCs.

given a weight that quantifies the probability of the gene being exclusively associated with the surrogate variables and not the outcome of interest. These weights are then used in the SVD calculation with higher weights given to rows not associated with the outcome of interest and associated with batches. Below is a demonstration of two iterations. The three images are the data multiplied by the weight (for a subset of genes), the weights, and the estimated first factor (code not shown).

```
## Loading required package: mgcv
## Loading required package: nlme
## This is mgcv 1.8-10. For overview type 'help("mgcv-package")'.

## Number of significant surrogate variables is:  5
## Iteration (out of 1 ):1

## Number of significant surrogate variables is:  5
## Iteration (out of 2 ):1  2
```

The algorithm iterates this procedure several times (controlled by B argument) and returns an estimate of the surrogate variables, which are analogous to the hidden factors of factor analysis. To actually run SVA, we run the `sva` function. In this case, SVA picks the number of surrogate values or factors for us.

```
library(limma)
svafit <- sva(geneExpression,mod)

## Number of significant surrogate variables is:  5
## Iteration (out of 5 ):1  2  3  4  5

svaX<-model.matrix(~sex+svafit$sv)
lmfit <- lmFit(geneExpression,svaX)
tt<- lmfit$coef[,2]*sqrt(lmfit$df.residual)/(2*lmfit$sigma)
```

There is an improvement over previous approaches:

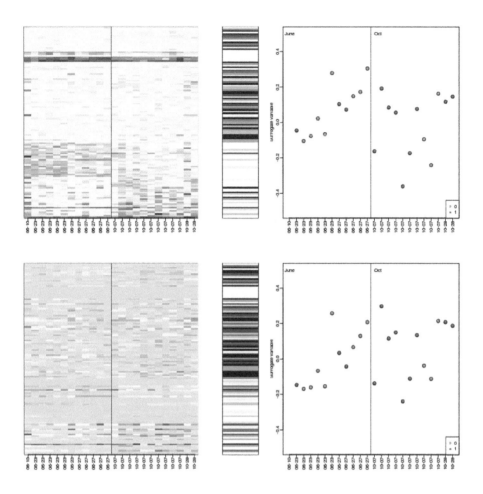

FIGURE 10.28
Illustration of iterative procedure used by SVA. Only two iterations are shown.

```
res <- data.frame(dm= -lmfit$coef[,2],
                  p.value=2*(1-pt(abs(tt),lmfit$df.residual[1]) ) )
mypar(1,2)
hist(res$p.value[which(!chr%in%c("chrX","chrY") )],main="",ylim=c(0,1300))

plot(res$dm,-log10(res$p.value))
points(res$dm[which(chr=="chrX")],-log10(res$p.value[which(chr=="chrX")]),col=1,pch=16)
points(res$dm[which(chr=="chrY")],-log10(res$p.value[which(chr=="chrY")]),col=2,pch=16,
       xlab="Effect size",ylab="-log10(p-value)")
legend("bottomright",c("chrX","chrY"),col=1:2,pch=16)

qvals <- qvalue(res$p.value)$qvalue
index <- which(qvals<0.1)

cat("Total genes with q-value < 0.1: ",length(index),"\n",
    "Number of selected genes on chrY: ", sum(chr[index]=="chrY",na.rm=TRUE),"\n",
    "Number of selected genes on chrX: ", sum(chr[index]=="chrX",na.rm=TRUE),sep="")

## Total genes with q-value < 0.1: 14
## Number of selected genes on chrY: 5
```

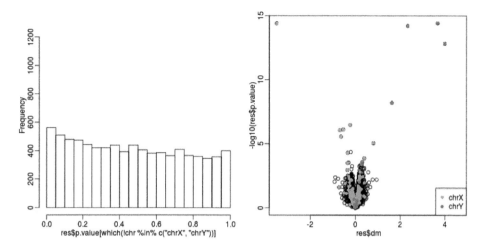

FIGURE 10.29
p-value histogram and volcano plot obtained with SVA.

```
## Number of selected genes on chrX: 8
```

To visualize what SVA achieved, below is a visualization of the original dataset decomposed into sex effects, surrogate variables, and independent noise estimated by the algorithm (code not shown):

FIGURE 10.30
Original data split into three sources of variability estimated by SVA: sex-related signal, surrogate-variable induced structure, and independent error.

10.13   Exercises

In this section we will use the `sva` function in the `sva` package (available from Bioconductor) and apply it to the following data:

```
library(sva)
library(Biobase)
library(GSE5859Subset)
data(GSE5859Subset)
```

1. In a previous section we estimated factors using PCA, but we noted that the first factor was correlated with our outcome of interest:

```
s <- svd(geneExpression-rowMeans(geneExpression))
cor(sampleInfo$group,s$v[,1])
```

The `svafit` function estimates factors, but downweighs the genes that appear to correlate with the outcome of interest. It also tries to estimate the number of factors and returns the estimated factors like this:

```
sex = sampleInfo$group
mod = model.matrix(~sex)
svafit = sva(geneExpression,mod)
head(svafit$sv)
```

The resulting estimated factors are not that different from the PCs.

```
for(i in 1:ncol(svafit$sv)){
  print( cor(s$v[,i],svafit$sv[,i]) )
  }
```

Now fit a linear model to each gene that instead of `month` includes these factors in the model. Use the `qvalue` function.
How many genes have q-value < 0.1?

2. How many of these genes are from chrY or chrX?

Index

A

Alternative hypothesis, 185

Analysis of variance (ANOVA), 160–166, 217
  CLT, 162
  contrast package, 164
  differences of differences when there is no intercept, 165
  different specification of the same model, 163
  estimated coefficients, 164
  F-distribution in, 217
  F value, 162
  least complex model, 161
  null hypothesis, 162
  original sum of squares, 161
  reduction in sum of squared residuals, 162
  residuals, calculation of, 161
  table, 161

ANOVA, *see* Analysis of variance

Association tests, 63–67
  Chi-squared test, 64
  Fisher's exact test, 64
  generalized linear models, 66
  hypergeometric distribution, 64
  "Lady Tasting Tea" experiment, 63
  odds ratio, confidence intervals for, 66
  p-value, 65
  sample size, p-values and, 66
  statistical question, 63
  two by two tables, 64

Asymptotic results, 10, 26

B

Bar plots, 79, 80, 81

Batch effects, 267, 305–342
  adjusting for batch effects with linear models, 325–329
    Combat, 327
    computing efficiency, 327
    error terms, 325
    improvement in specificity, 326
    model matrix, 325
    p-value histogram, 327
    t-test, 325
  confidence interval, construction of, 306
  confounding, 307–310
    average after stratifying, 309
    baseball, 310
    chi-square test, 307
    explained graphically, 308
    high-throughput example, 311–312
    Simpsons paradox, 307, 310
    stratified data, 309
    volcano plots, 312
  discovering batch effects with EDA, 313
  estimate of spread, 305
  exercises, 312–313, 321–322, 329–330, 333–334, 342
  factor analysis, 330–333
    correlation structure, 332
    example, 330
    factor model, 331
    images of correlations, 331, 333
    PCA, 332
    sample correlations, 330
    STEM subjects, 331
  factor analysis, modeling batch effects with, 335–341
    batch effect, removal of, 337
    data set, 335
    dates with color denoting, 336
    PCs stratified by dates, 338
    p-value histogram, 339
    subset gene expression data, 335
    Surrogate Variable Analysis, 338, 340
    variance explained, 337
  gene expression data, 314–321
    boxplot of PCs, 318, 320
    calculating the PCs, 315
    detrended gene expression data matrix, 314
    exploratory plot, 316